UNIVERSITY OF WINNIPEG
LIBRARY
DISCARDED

PHILOSOPHY OF TECHNOLOGY

PHILOSOPHY AND TECHNOLOGY

Series Editor: PAUL T. DURBIN

Editorial Board

Albert Borgmann, *Montana*
Mario Bunge, *McGill*
Edmund F. Byrne, *Indiana – Purdue at Indianapolis*
Stanley Carpenter, *Georgia Tech*
Robert S. Cohen, *Boston*
Ruth Schwartz Cowan, *SUNY – Stony Brook*
Hubert L. Dreyfus, *California – Berkeley*
Bernard L. Gendron, *Wisconsin – Milwaukee*
Ronald Giere, *Minnesota*
Steven L. Goldman, *Lehigh*
Virginia Held, *CUNY*
Gilbert Hottois, *Université Libre de Bruxelles*
Don Ihde, *SUNY – Stony Brook*
Melvin Kranzberg, *Georgia Tech*
Douglas MacLean, *Maryland*
Joseph Margolis, *Temple*
Robert McGinn, *Stanford*
Alex Michalos, *Guelph*
Carl Mitcham, *Polytechnic University*
Joseph Pitt, *Virginia Polytechnic*
Friedrich Rapp, *Dortmund*
Nicholas Rescher, *Pittsburgh*
Egbert Schuurman, *Technical University of Delft*
Kristin Shrader-Frechette, *South Florida*
Elisabeth Ströker, *Cologne*
Ladislav Tondl, *Czechoslovakia*
Marx Wartofsky, *CUNY*
Caroline Whitbeck, *M.I.T.*
Langdon Winner, *R.P.I.*
Walther Ch. Zimmerli, *Technical University Carolo-Wilhelmina, Braunschweig*

OFFICIAL PUBLICATION OF
THE SOCIETY FOR PHILOSOPHY AND TECHNOLOGY

PHILOSOPHY AND TECHNOLOGY
VOLUME 6

PHILOSOPHY OF TECHNOLOGY

Practical, Historical and Other Dimensions

Edited by

PAUL T. DURBIN

University of Delaware

KLUWER ACADEMIC PUBLISHERS

DORDRECHT / BOSTON / LONDON

Library of Congress Cataloging-in-Publication Data

Philosophy of technology: practical, historical, and other dimensions /edited
   by Paul T. Durbin.
   p. cm. (Philosophy and technology: v. 6)
Includes index.
ISBN 0-7923-0139-0.
1. Technology–Philosophy.   I.   Durbin, Paul T.   II.   Series.
T14.P53   1989
601–dc 19   89-2765

Published by Kluwer Academic Publishers,
P.O. Box 17, 3300 AA Dordrecht, The Netherlands.

Kluwer Academic Publishers incorporates
the publishing programmes of
D. Reidel, Martinus Nijhoff, Dr W. Junk and MTP Press.

Sold and distributed in the U.S.A. and Canada
by Kluwer Academic Publishers Group,
101 Philip Drive, Norwell, MA 02061, U.S.A.

In all other countries, sold and distributed
by Kluwer Academic Publishers,
P.O. Box 322, 3300 AH Dordrecht, The Netherlands.

printed on acid free paper

All Rights Reserved
© 1989 by Kluwer Academic Publishers
and copyrightholders as specified within.
No part of the material protected by this copyright notice may be reproduced or
utilized in any form or by any means, electronic or mechanical,
including photocopying, recording or by any information storage and
retrieval system, without written permission from the copyright owner.

Printed in the Netherlands

TABLE OF CONTENTS

PREFACE vii

Introduction: General Perspectives on the Complexity of Philosophy of Technology, Friedrich Rapp ix

PART I
PRACTICAL PROBLEMS

LYLE V. ANDERSON / Cybernetics, Culpability, and Risk: Automatic Launch and Accidental War 3

DAVID BELLA / Catastrophic Possibilities of Space-Based Defense 27

EDWIN LEVY / Judgment and Policy: The Two-Step in Mandated Science and Technology 41

PART II
HISTORICAL DIMENSIONS

LEE W. BAILEY / Skull's Darkroom: The *Camera Obscura* and Subjectivity 63

EDMUND F. BYRNE / Workplace Democracy for Teachers: John Dewey's Contribution 81

LARRY HICKMAN / Doing and Making in a Democracy: Dewey's Experience of Technology 97

JOSEPH MARGOLIS / Pragmatism, *Praxis*, and the Technological 113

PART III
INTERNATIONAL AND INTERGENERATIONAL PERSPECTIVES

GAO DASHENG and ZOU TSING / Philosophy of Technology in China ... 133

WOJCIECH GASPARSKI / Design Methodology: A Personal Statement ... 153

JANET FARRELL SMITH / Responsibility and Future Generations: A Constructivist Model ... 169

Name Index ... 187

PREFACE

The corps of philosophers who make up the Society for Philosophy & Technology has now been collaborating, in one fashion or another, for almost fifteen years. In addition, the number of philosophers, world-wide, who have begun to focus their analytical skills on technology and related social problems grows increasingly every year. (It would certainly swell the ranks if all of them joined the Society!) It seems more than appropriate, in this context, to publish a miscellaneous volume that emphasizes the extraordinary range and diversity of contemporary contributions to the philosophical understanding of the exceedingly complex phenomenon that is modern technology. My thanks, once again, to the anonymous referees who do so much to maintain standards for the series. And thanks also to the secretaries – Mary Imperatore and Dorothy Milsom – in the Philosophy Department at the University of Delaware; their typing and retyping of the MSS, and especially notes and references, also contributes to keeping our standards high.

PAUL T. DURBIN

FRIEDRICH RAPP

INTRODUCTION: GENERAL PERSPECTIVES ON THE COMPLEXITY OF PHILOSOPHY OF TECHNOLOGY

Any general programmatic definition must by its very nature remain abstract and vague, and defining the philosophy of technology is no exception to this rule. Nonetheless, there should be no objection to this almost tautological statement: The philosophy of technology investigates the historical and systematic philosophical problems raised by the development of technology – though even on this there is not likely to be universal consent. What obtains here applies to all general declarations, be they about humanism, civil rights, or the preservation of peace. In all such cases, while it is easy to arrive at a generally accepted formula, the formula will unavoidably be so vague that everybody is free to find his or her own interpretation.

Conversely, if a notion has been worked out in detail, it can be discussed in a concrete way, and people can arrive at consensus or disagreement. The real problems arise when one supplies a more specific notion of the philosophy of technology. As in any other field of scholarship, it is virtually impossible to elaborate a philosophy of technology *ex nihilo*. One must draw, explicitly or implicitly, on the given state of the art. One may in fact change or even negate this state, but in any case it must serve as a base for further development. But what is lacking in the philosophy of technology is precisely a well elaborated state of the art. The situation is different from other fields of philosophical inquiry. In such areas as the philosophy of history, ethics, social philosophy, political philosophy, phenomenology, philosophy of language, or philosophy of science, there have been long standing discussions; there is a well established, systematic conceptual framework of basic concepts, questions, theses, and arguments. As a result, in these fields there is a clearly established level of scholarly discussion. For the philosophy of technology, a similarly detailed and elaborate theoretical frame of reference is mainly a desideratum. The field is still in the making.

WHY TECHNOLOGY HAS BEEN NEGLECTED IN PHILOSOPHY

Clearly this neglect in philosophical research is not caused by the lack of relevance of technology. Just the contrary. It is a commonplace that modern science and technology, and the process of industrialization they enhance – together with secularization, enlightenment, democracy, equal rights movements, and bureaucratization – are the leading forces of a time often called 'the age of technology.'

Four reasons can be adduced why technology has been neglected – and in fact is still largely neglected – by philosophers. Two of the reasons originate from the history of ideas; the other two are of a pragmatic nature.

(1) Since its beginnings in antiquity, the Western history of ideas has been dominated by the notion of theory. Dealing with theoretical questions is deemed a superior form of life, ranking well above mere practical activity or craft work. In these terms and up to the present, technology is given a lower theoretical and philosophical ranking. The 'two-cultures' debate initiated by C.P. Snow shows that this split remains relevant today.

(2) The second theoretical reason is a false notion, that technology is nothing but applied science. If this notion is taken for granted, it follows that there is no reason for a special philosophy of technology; the philosophy of science will automatically cover all relevant problems. Yet this approach is clearly mistaken. There is indeed a close interrelation between science and technology. Every effect produced in the laboratory is by its very nature a technological phenomenon, and every scientific finding can be used to bring about technical effects. To put the matter bluntly, modern science is technology – as far as experimental procedures are concerned. But scientific research does not by itself or easily translate into technical practice. To make the transition, all the skill and knowledge of the engineering arts and sciences are needed. Clearly the structure of thinking in the technological sciences, as well as the methodological principles of design and of efficient and purposeful action, exhibit patterns of their own. So they can by no means be reduced to the philosophy of science.

(3) Furthermore, there is a practical reason for the fact that only in recent times has technology attracted much attention. Before the time of the Industrial Revolution, analogously to the biological features of humankind, technology was just taken for granted; it was not regarded as deserving special theoretical investigation. This remained true with Hegel, Burckhardt, and Dilthey. The notion that the development of technology is a decisive factor in history originated only at the time when technological change became highly visible – i.e., with Adam Smith (1723–1790), Saint-Simon (1760–1825), and Karl Marx (1818–1883). From the point of view of epistemology, the interpretation of historical change in terms of technological development is just a projection backwards of experiences after the Industrial

Revolution.

This is related to the first reason, above. It is only by means of the objectivizing, reflexive, theoretical type of thinking which is characteristic of the Western tradition that the detached, neutral analysis of given states of affairs characteristic of modern science and technology came into being. Whereas in craft work one aims immediately and directly at achieving some desired result, in technology detours are deliberately made and complex processes are brought into being so that the goal desired can be better attained by means of mechanical, electrical, or chemical processes. The irony is that the type of thinking which gives technology an inferior ranking is what brought about modern technology. In other words, modern science-based technology got its origin from the Western theory-oriented attitude, even though it is so often regarded as an illegitimate offspring.

(4) Finally, the fact that philosophical investigations of technology are only of recent origin can be ascribed to the complex, multidimensional features which technology exhibits. All the scientific disciplines and the different fields of philosophy have their origin in the methodological procedure of abstracting specific features and investigating the theoretical structures thus obtained in systematic, coherent models. But technology, along with economics, politics, and history, is made up of a variety of closely connected, mutually dependent phenomena. As a result, it is only at the price of oversimplification that one can formulate structural models or handy theories. However, it is precisely the totality of the interrelations which is characteristic of modern technology; so, strictly speaking, any type of simplification is inappropriate. For this reason, working out a philosophy of technology is a fascinating but at the same time a very complicated endeavor.

DIFFERENT APPROACHES

In dealing with the complex phenomenon of technology, two extreme approaches can be distinguished. They are like Weberian 'ideal types' intended to reduce the holistic phenomenon to manageable conceptual models. One approach takes the totality as a basic concept that cannot be broken down to simple elements; according to this approach, the totality is open to description and explication only as a whole. This strategy, reminiscent of Hegel, is chosen by Ellul and Heidegger – the latter on a metaphysical level. The other approach introduces analytical distinctions, interrelations, and dependencies which explain how the phenomena have come about. In the first approach the totality is dealt with as a sort of black box, without considering the elements and their interrelations. The second approach conveys insights about details but may miss the whole. Yet, analytical method has the advantage of highlighting specific, intelligible relationships that otherwise might remain unidentified. This has the further advantage of showing how these features can be changed. In contrast to this, the totality approach has to accept the process of technological change as a given that can not be further analyzed. The analytic approach provides a way to consider, alongside *a priori* philosophical arguments, the contingent *a posteriori* historical, cultural, and social factors relevant to the development of technology.

In fact, both the process which brought about technology and the concrete features which technology exhibits are contingent historical phenomena. This is why the philosophy of technology must necessarily consider empirical facts, just as a philosophy of history that wants to go beyond mere speculation must be based on the conceptual reconstruction of the real past. But turning to empirical findings also involves a danger. If the philosophy of technology were to aim exclusively at giving a popularized, synoptic account of empirical findings, the philosophical dimension would be lost; by definition, philosophy transcends the realm of contingent, empirical facts. This is a special case of the notorious dilemma, between taking the abundance of empirical findings for granted and capricious philosophical speculation. This dilemma, inherent in modern philosophy, is intensified in the case of the philosophy of technology. Taking technology in its totality, there are myriad scientific disciplines to be considered: history, the technological sciences, the natural sciences, economics, political science, sociology, psychology; moreover, different areas of philosophy are needed: philosophy of history, ethics, theory of knowledge.

THE CONTENT OF EACH APPROACH

It is only in abstraction that the appropriate methodological procedure can be separated from the contents of the philosophy of technology. Here also two extreme cases can be distinguished and the choice will unavoidably establish a precedent for the philosophical results one can arrive at. In the broader definition, technology comprises any systematic, goal-directed procedures that go beyond unstructured, elementary actions. In these terms, not only complex activities such as space technology and energy supply – which involve different steps, the division of labor, and concerted action – but even the method of playing football or the piano must count as belonging to the field of technique and thus of technology. A narrower, more limited understanding of technology is involved if one takes modern technology as the point of reference. In the latter case it is the results, the concrete artifacts of the material technology by which man uses the forces of nature for his own purposes, that count. The fact that good reasons for both types of definition can be abused gives witness to the close connection between procedures and material systems. Once again, this indicates the broad range of the phenomenon of technology, and this cannot be eliminated by terminological stipulations. It seems that the second, the more concrete notion of technology, is of greater heuristic use when it comes to understanding and analyzing the large-scale transformation of the physical world by modern technology. However, in contrast to this, the broader notion of technology is more appropriate when it comes to dealing with the different types of techniques and tools that have been applied in various historical epochs.

In a similar vein, the economic rationalism, the efficient management of modern life described by Max Weber, is closely related to the state of material technology. In the industrialized nations we everywhere meet with technological artifacts that have turned into a sort of second nature. To a large degree, we can no longer survive without these auxiliaries. The result is that today we no longer fight primarily against the hostile forces of nature. However, this does not mean that we have attained the 'world of freedom' that was optimistically expected in the Age of Reason and in the nineteenth century. The constraints of nature have largely been replaced by man-made technological constraints and by unintended secondary effects such as ecological problems and the scarcity of resources.

If we want to profit from technological processes and systems, we are unavoidably forced to adjust more or less to their functioning principles, such as the division of labor, standardization, accuracy, shift work. We have to pay a price for the services technology yields. Of course it is possible to soften the effects of these technology-included constraints. But they cannot be abolished in principle since they arise from functional relationships and from the laws of the physical world. Here Francis Bacon's phrase, *Natura non nisi parendo vincitur*, applies; technological action is based on obedience to nature. Adjusting to monotonous, mechanical, inhumane processes, losing the immediate contact with a style of life close to nature and to those organic rhythms to which we are subject as creatures of nature – all this unavoidably implies an alienation from our natural origins. It is only on holiday, in leisure-time activities, that city dwellers can realize a life close to nature if they desire such.

Mechanical processes most often follow the lines of the rotating wheel. Constant, monotonous rotation is an inorganic principle which has no analogue in organic processes. As compared to the differentiated and variable processes of life such monotonous processes are rigid and inflexible. Mechanical processes are, more than anything else, designed to substitute for or to enlarge the forces of human muscles. In contrast to this, electronic control systems are closer to the human nervous system, and in principle they offer a better chance that technology can be adjusted to humans rather than the other way around. In this perspective, computer technology may contribute to making technology more humane. Yet these positive characteristics are inherently connected with negative possibilities. There is a clear danger in information technology that language, the most important organ of culture, will be reduced to standardized abbreviations and preset structures and categories. Everything that does not fit within these schemes might fall victim to technological selection. This has happened in television which tends to reduce variety, creativity, and intellectual spontaneity because the viewer is merely watching pictures and not applying conceptual structures and theoretical reasoning, as he or she would in reading a book.

From these examples we learn how technology is not only different from and opposed to culture but at the same time an element of culture. The way technology is brought into being and put to use is determined by the values cherished in a culture. Yet, being a decisive factor of modern life – of economics, politics, and culture – technology in its turn will inevitably also shape the priorities taken for granted in these fields. How could it be different? Technology produces the material framework within which we live our daily lives individually and collectively.

Thus technology inevitably affects the way in which we understand the world and ourselves. The result is that in private life as well as in professional life, humans are to a large degree considered in machine terms. Technology occupies a place formerly held by politics, art and religion. Technological expressions, such as 'shifting,' 'cranking up,' 'idling,' and 'sand in the works,' have become a part of everyday speech. In the history of ideas, the close relationships between technology and the image of man become evident in Descartes's interpretation of man as automaton, in Hobbes's mechanistic theory of the state, and in Freud's notion of pent up psychic energy. More recently, communication theory is based on the paradigm of transmitter and receiver; and cognitive psychology is shaped according to the conceptual framework of information technology.

THE NECESSITY OF TECHNOLOGY

Surely nobody would seriously want to abolish modern technology because of such problems. Indeed, this would be impossible since we could not even survive, let alone maintain our standard of living, without making use of science and technology. In methodological terms, technology has an instrumental function; it is used to attain a given goal with the least possible effort. It is in these terms that choices among competing technological solutions have always been made. The choice is accepted by society, at least *ex post facto,* since otherwise the technology in question would not survive. This implies that the present state of technology did come about *because* it was accepted by the majority of the population. Even an 'alternative,' 'soft,' or 'appropriate' technology that fits the situation of developing countries would surely make use of the latest state of the skill and knowledge available.

The limitations imposed upon political and economic activity by technological constraints – as discussed often under the heading, 'the technological state' – are, strictly speaking, only conditional. In a way similar to the laws of nature, these constraints are operative only under a proviso, that certain conditions are brought about, or that specific results are desired. If there are no other means for attaining these goals and if one wants unconditionally to attain these goals, the constraints in question are unavoidable. Here the real, physical world offers resistance, acts as an obstacle, which cannot be removed by normative postulates or acts of will. In historical terms, the needs, values, priorities, or goals which direct the activity of an individual – or of social groups, nations, even mankind as a whole – are not predetermined biologically. As soon as questions are raised that go beyond the mere physical maintenance of life, there is always an element of cultural choice in play which is to a large degree predetermined by the given historical tradition.

We witness an ever accelerating pace of technological change. This is based on the momentum of accumulated technology as well as on the exhaustion of the potentials of instrumental reasoning by way of specialized research and planned development. It is, in other words, a result of 'the invention of invention' (Whitehead). In philosophical terms, such rapid development is a *contingent* phenomenon; it is neither a law of nature nor a product of historical necessity. This is why in each situation, at least tacitly, a cost-benefit analysis must be made in order to arrive at a social consensus about the advantages and desirability of the proposed technology, along with the disadvantages one will have to accept. This process of social choice also involves economic, political, social, and cultural systems, all of them acting together in a complex, interrelated process.

MATERIALIST AND IDEALIST DIMENSIONS

There are good reasons for a materialist interpretation of technology, i.e., for technological determinism; but also for an idealist interpretation, i.e., for value determinism. Applying another wording, one can distinguish between real and ideal factors here. The materialist interpretation refers to the quasi-institutionalized, apparently self-sustaining process of technological development. By means of highly specialized basic research which relies on the accumulated fund of technological and scientific findings and is fostered by international competition, ever new, unexpected technological possibilities are provided. By means of subsequent economic filters, certain of these possibilities are selected for further development and finally put into practice. In terms of this model, cultural change will follow, after a certain lag, changes in material technology.

Clearly there is momentum in the state of affairs brought about by earlier choices and existing trends. This notwithstanding, the process is subject to the free and spontaneous choice of the actors involved. For this reason, it is possible to make an appropriate re-evaluation – for example to decide to protect the natural environment or to preserve limited resources – and set new priorities for the future so that other lines of development can be followed. The 'superstructure' of the closely-knit web of science, technology, and industry is a firmly fixed element of contemporary life. But it does not exist by itself. It was brought about by man. Both tacit consent and activity corresponding to that consent are needed for that superstructure to continue to exist in the future. So it turns out that dichotomized models – which describe technology either as a natural process or as a cultural phenomenon, as determined either by real or by ideal factors, as materialist or idealist – are no more than oversimplifications intended to draw our attention to certain features of the situation.

In actual fact, it is necessarily the case that both elements always come into play. In most general terms, technology consists in the shaping of the physical world according to human purposes. Hence it is at the same time a phenomenon of both nature and culture. The contrariety but also the unity of these two poles affects the nature of human nature itself: we are at the same time physical and mental beings. We have a body but we are also capable of thinking, willing, and feeling. In Descartes's terms, a human being is *res cogitans* and *res extensa* at the same time. Our participation in the ideal as well as in the material sphere is, in the last analysis, the reason for the bipolar structure of technology. In a biological perspective, *Homo sapiens* has an inadequate endowment. Our species needs appropriate tools to compensate for deficiencies in human organs to ensure survival in a hostile environment. The functional substitute for this inadequate endowment consists in creative problem-solving using appropriate tools: man is the tool-making animal (Benjamin Franklin), or *Homo faber* (Bergson).

The first impulse toward technology is natural. Following this impulse, humans were no longer tied to the immediate concrete situation; they managed to transcend the given by means of intellectual or theoretical variations. This started the way toward detachment from the immediately given, toward objectivizing reflection. But the present state of development, brought about by sophisticated technological and scientific methods, goes far beyond the compensation of an inadequate biological endowment. After all, until very recently and perhaps even today, there have been people living at the technological level of the stone age. And it is not nature but modern man that threatens them with extinction. The conclusion? A purely naturalistic or materialistic explanation of the development of technology cannot do full justice to historical development as it actually occurred.

CULTURAL PERSPECTIVES

When it comes to explaining the dynamics of technological change it is not enough merely to refer to the perfection of technological means, to systematic research, to the accumulation of technological skill and knowledge during recent centuries, or to the interrelation between science, technology, and industry. All of these factors, taken singly or in conjunction, cannot give an exhaustive explanation of past processes, or of the present situation. Nor can what has happened be convincingly explained by the human needs or political interests to be fulfilled by means of technology. As Ortega y Gasset has said, being human means being able to say No. In every culture, the ascetic renunciation of satisfying even apparently vital needs is honored as giving witness to high ideals. However, surely technological possibilities and human needs do promote each other. And as soon as the fulfillment of needs is within reach, an act of renunciation, whether forced or self-imposed, is regarded as a lack of individual or social self-fulfillment. Once needs go beyond elementary biological necessities for the preservation of life, they are shaped by culture. For this reason, we must regard the passive acceptance or the active fostering of the trend toward increasing technology as a cultural reaction – that is, as affirming a certain value system. In the last analysis, this must be seen as an act of will. This 'act of technological will' manifests itself in the Promethean feeling of modern times, in creative activity, in technological production, in the striving for power, in the aim to overcome space, time, and even death. In philosophical terms, this can be interpreted as a collective effort to transcend the limits of finite existence.

This is not to say that technological systems are in a unique way tied to specific cultural value systems. Since they enlarge and extend biological nature in a way that is common to all and which is easily perceived by everyone, technological results, once achieved, are accepted world-wide, independent of the given cultural and historical traditions. All technological systems and processes, directly or indirectly, extend our physical capacities, either sensuous or calculative (in the case of computers). These increased capacities can be used only if they are in one form or another accessible to the senses. This is evident in such fields as transportation technology (which aims at moving things around bodily), in information technology (where the manipulated symbols must be visible or otherwise clearly identifiable), and unfortunately also in the visible destructive power of military technology. In fulfilling given functions the superiority of modern technology is immediately apparent. This should not be astonishing, since a given choice among possible technological solutions would have been made because the function in question would thereby be carried out most efficiently and completely.

What technology aims at can only be attained, indeed, when technological systems and processes are run efficiently. For this reason, the 'software,' the attitude appropriate for implementing material technology, must be available.

This implies that an appropriate infrastructure, including the ability to do the required work in a disciplined manner, is presupposed; without these elements, there is no way of efficiently using the technological 'hardware.' Thus at least a certain degree of the Western, technology-oriented style of life must be imported along with and as part of technology. A successful transfer of technology is bound to a corresponding transfer of culture.

In industrialized nations, with their history of technological development, there is fertile ground for efficiently putting technological systems to use. The transfer of culture which is necessary in order to run modern technology efficiently in developing countries gives rise to various problems. It is a question to what degree it is possible to profit from modern technology, integrating it efficiently into a native cultural tradition. There is a danger of losing the individual and collective identity shaped over years of history and rooted in that culture's value system.

It is worth noting in passing that, in this process, expansionist Western thinking may be stimulated reciprocally by impulses from other cultures. A monolithic global culture exclusively oriented toward scientific, technological, and economic efficiency would destroy the variety of existing cultures and value systems. Such a development would clearly result in a cultural impoverishment of humankind.

A PARADOX: SUCCESSFUL TECHNOLOGY BREEDS SOCIAL PROBLEMS

It is a paradox that many of the problems raised by modern technology arise not from failures but from too much success. Since the Industrial Revolution – which in its beginnings was based on improvements in organization and craftsmanship rather than on scientific principles – an ever closer interrelationship between science and technology gradually developed. In this way, unexpected results were attained in all areas of life; and there is no end in sight. Along with the results aimed at, often unintended and unexpected secondary effects have increased, too. Technology, designed to replace manpower and to raise productivity, to the degree that this goal is achieved raises problems of unemployment and of the proper use of leisure time. In Europe and in North America, modern agricultural technology has turned out to be too successful: farmers suffer from surplus production. Pushing for higher standards in housing conditions leads to uncontrolled urban sprawl. The passenger car provides greater mobility, but also noise, air pollution, and the despoliation of the countryside. The progress of medicine, so closely connected with technological progress, has increased life expectancy, but at the same time this progress resulted in the population explosion in developing countries. Even in industrialized nations, the question is discussed as to what degree the prolongation of life at any price conflicts with human dignity.

In many cases, such problems are not completely new but aggravations of earlier difficulties. Throughout history, natural resources have been exploited, the countryside has been despoiled (e.g., the deforestation of Mediterranean countries), and the latest developments of technology have been used for military purposes. However, due to the very success of modern technology today, the problematic effects have increased so much in scale that there has been, as it were, a qualitative change: entirely new dimensions of problems have arisen. Strictly speaking, limits of natural resources and the apocalyptic power of atomic weapons call for voluntary global restrictions – including international agreement and cooperation – not for growth and power without restraint. Considering the far-reaching, virtually irreversible effects, we need to act with a high degree of responsibility. Yet a global ethics – reaching out in space and time, considering both the interests of other parts of the globe and the prospects for generations to come – apparently surpasses our intellectual and moral capacities, as well as the always restricted potentials of human activity. There is an exact correspondence between the range within which we can make reasonable predictions or within which we can control natural and social processes, on the one hand, and the range of our sense of responsibility. In either case only the short range is covered. Predictability, control, and responsibility for distant phenomena and processes fade away very quickly.

PROBLEMS WITH ETHICAL MAXIMS FOR TECHNOLOGICAL CONTROL

A similar difficulty obtains with respect to the concrete maxims of an ethics of technology. The individual and society develop by realizing their potential. But modern technology has increased the range of our potentialities to such a degree that the maxim can no longer be, 'Develop your potentialities!' Now it must be, 'Avoid doing all you can do!' Already today, not to speak of the future, limited resources – of labor, capital, machinery, raw materials – have given rise to a selection process mandated by economics. This is to say that within the virtually infinite realm of what *could* be done, in terms of technological skill and knowledge, only a very limited part can actually be realized. For example, only a very small percentage of patents granted are actually put to use. However, up to now this principle of selection has applied only to concrete details, not to technological principles. Humankind has not yet refused to step into the new dimensions opened up by the progress of science and technology.

Even in such controversial fields as biomedicine and genetic engineering, a good case might be made for doing everything we can do. However, the lesson of history is that, once the first step has been taken, development cannot be stopped; global scientific, technological, and economic systems of exchange and competition guarantee that whatever has become standard will very soon spread all over the world.

For this reason, it is easier and safer to stop before getting started than it is to slow development already in progress. For example, in genetic engineering, the nature of the human individual, at least the physical side, is in question. Therefore, a discussion of fundamentals is needed to find out how far one can go without putting at risk human dignity and the unique character of each individual person. Clearly, this example is critical. Situations of this type can only be handled successfully if technological possibilities are matched by an appropriate understanding of the problem and by an adequate sense of responsibility. It is a positive sign that, in recent years throughout the world – though in varying degrees – sensitivity to these problems has increased. It is also a positive sign that the new consciousness arose from a lower threshold of tolerance, and not so much from an aggravation of the problems in question. Since this values change took place in a rather short time, one may hope that humankind will succeed in time in finding new norms to deal reasonably with the dynamics of scientific and technological change.

TECHNOLOGY ASSESSMENT

In order to resolve issues of this sort, arguments against certain technological innovations may need to be discussed by the broad public in order to arrive by rational choice at a generally accepted consensus. This presupposes a forecast of the results to be expected from the different alternatives at hand, as well as an evaluation of their desirability. In doing this, the context should require more than the criteria of technological feasibility and market acceptance. In broadening the frame of reference, both the expected and desired and the unexpected and undesired consequences – political, social, ecological, and cultural – should be taken into account. This is the process commonly called 'technology assessment.' In its details, it is open to different interpretations: it is a means for projecting alternative trends.

In general terms, technology assessment relies on ideas from the Enlightenment, on rational decision theory, and on technological procedures. This is to say that, here again, as with such concrete problems as protecting the natural environment or preserving natural resources, the problems raised by technology can only be solved by more technology (perhaps of a different type) – not by renouncing technology altogether. After all, deliberate planning, the forecasting of consequences to be expected, choosing the appropriate alternative, and implementing the solution chosen – all of these are inherent features of technological action. In highly speculative terms, this is reminiscent of the theological dictum, *Nemo contra deum nisi deus ipse* ('Only God can counter God').

Here, the close link between technological action and rational procedures becomes evident. In fact, the only alternative to a sober analysis – in which prognoses and evaluative assumptions are made explicit and laid open to examination – would be intuitive decision making based on pure guesswork.

RATIONAL DECISION PROCEDURES

Rational choice procedure requires the assessment of alternatives, and, much as it is in need of improvements, we must rely on this procedure. However, we must also keep in mind that there are basic limits to rational decision making; in the last analysis, these limits arise from the imperfect human condition. The technological results of a certain course of action can be predicted with a rather high probability; this is why engineering and technology succeed so often in accomplishing their goals. But the far-reaching ecological, political, social, and cultural consequences of any particular innovation can only be predicted with subjective probability. Different people, and even different experts, may arrive at divergent results. This is not surprising. In the wealth of its concrete consequences, technology is no exception to the rule that, in history, people do not really know what they are doing; the real consequences of their actions become evident only in hindsight.

Realizing the limits of rational decision making should not suggest that we should refrain from making forecasts or evaluations. In retrospect, considering the expectations that contemporaries had about the technological innovations of earlier times (the automobile, airplanes, radio broadcasting), it becomes evident that the full range of the consequences that did actually arise was by no means known to them. There is no way of avoiding unforeseen consequences, but we can and must do our best to keep these as small as possible.

THE VALUE DIMENSION

Even if, contrary to fact, we assume that all the consequences were known, the problem of evaluation would still be there. A choice among alternatives had to be made. Since such decisions concern the whole of society, economic and political theory inevitably come into play here. And what we need is a normative conceptual model of the complex, interrelated decision processes involved. Here philosophy is, above all, concerned with the values and priorities in terms of which a choice is to be made – that is, it is concerned with ideal notions. For the individual, the ideal of a good life and of a humane existence act as the ultimate point of reference; for society, justice, equality, and the common good are the ultimates. In historical perspective, the goal or telos of history itself acts as the ultimate point of reference.

Surely one cannot expect to solve these deep philosophical problems with *ad hoc* solutions prepared just to complete a philosophy of technology. The philosophy of technology here must rely on broader philosophical discussions. There is no possibility of developing solutions in isolation that would go farther than what has been achieved in the field as a whole. Since almost all discussions of social theory, of ethics, of the philosophy of history are controversial, one cannot reasonably expect in the philosophy of technology to arrive at perfect and undisputed solutions. On the other hand, it is also true that one cannot wait until these broader philosophical disputes are settled. One must be prepared to put up with provisional solutions, with the best one can achieve for the time being. It is a positive feature of the philosophy of technology that it keeps reminding us of the practical relevance of the problems discussed. This opens up the possibility that traditional philosophers will turn to these problems, recognizing a new field of research – and thus that traditional philosophy will be taken more seriously.

CONCLUSION

Summing up, we can say that the ambivalence of technology described in mythology in the figures of Prometheus and Icarus belongs to its very essence. It cannot be abolished. Since in simplest terms, technology is the reshaping of the physical world for human purposes, and since this reshaping results in concrete technological systems and processes which create a new type of reality, humans will have to adjust. Technology facilitates and liberates, but in doing so it also creates new burdens and constraints. Since the forces of nature are, as it were, blind with respect to the use we make of them, large-scale production necessarily also implies large-scale destructive power. Civilian and military uses of atomic energy are examples of this. Technology, hailed as a guarantee of human development and social progress, still exhibits inhumane and destructive powers. In utopian visions, technology is celebrated as a means of liberation, but there are also dystopian writings that describe the technological manipulation of the human personality, turning it into something purely mechanical, void of intellect and soul.

Those who welcome modern technology can rightly claim that it is a genuine continuation of the Renaissance, of the ages of Rationalism and Enlightenment. Critics with an artistic, romantic, or existentialist background, as well as social critics, complain that modern technology gives rise to alienation, consumerism, and the despoliation of nature. In religious terms, technology appears simultaneously as a curse and a blessing. Surely all of these statements are exaggerated. Technology, the center of modern life, is being taken as a focus for all the problems of the modern world. However, the crisis aspect of recent assessments of technology suggests that we have moved beyond superficial concerns with mere 'negative externalities' to a deeper awareness of the inherent ambivalence of technology.

Since philosophy contributed to unchaining the dynamics of modern technology, it should be capable of helping to clarify our present situation and of helping to guide technological development in reasonable directions.

University of Dortmund

NOTE

This introduction was prepared especially for this volume, although for the most part it is an edited and enlarged version of my 'Perspektiven einer Philosophie der Technik' (*Revue Internationale de Philosophie*, 161 [1987]:171–183). For another survey, with extensive references, see my 'Philosophy of Technology,' in G. Fløistad, ed., *Contemporary Philosophy: A New Survey*, vol. 2: *Philosophy of Science* (The Hague: Nijhoff, 1982), pp. 361–412. A condensed version of the latter survey, again with extensive references, appeared as 'Philosophy of Technology: A Review,' in *Interdisciplinary Science Reviews* 10 (1985):126–139.

PART I

PRACTICAL PROBLEMS

LYLE V. ANDERSON

CYBERNETICS, CULPABILITY, AND RISK: AUTOMATIC LAUNCH AND ACCIDENTAL WAR*

I

In this paper I will pursue an interesting if frightening isomorphism between Aristotelian and Marxian paradigms of deception. The former is intrapersonal: a self-deceived agent devises strategies to prevent himself from knowing either a principle or a consequence of some action that would make it immoral to perform that action.[1] The latter paradigm is interpersonal: a political deceiver is one who tries to prevent others from knowing either the principles or consequences of his own actions. Physicist Ray Kidder of Livermore Labs, who resigned their Star Wars project, says of this latter case that 'the public is getting swindled by one side that has access to classified information and can say whatever it wants ..., whereas [skeptics] ... would go to jail [because of Reagan's 'gag rule' on research].'[2] A political decision structure can be *designed* to preclude public discussion of the projectible consequences of particular programs or policies. Paradoxically, then, as outsiders we must devise independent means to discover these consequences in order even to raise this question of deception, and thereby to address the issue of culpability in the case of 'accidental' war.[3] Luckily, such independent means exist, in the accidental war studies conducted by mostly former military personnel who preceded the governmental gag rule on military research.

There is little precedent in the 'strategic' literature regarding the structure of this deception relation. John Steinbruner[4] has argued that there is a 'cybernetic' relation between the intended and the actual consequences of policy decisions. By virtue of potentially fatal flaws in the bureaucratic structure — which has taken on a life of its own after its designer (*kubernetes*, 'steersman') has set it in motion — strategic military decisions can produce exactly the opposite effect of the policy intended. Steinbruner has impressively analyzed the structure of our 'nuclear sharing' decision with NATO allies, to show how our intention to create a limited multilateral nuclear force (MLF) has unintentionally resulted in multilateral proliferation.

But, for two related reasons, I think that Steinbruner's analysis is insufficient. First, let us look at the nuclear analogy to the classic cybernetic relation

Paul A. Durbin (ed.), Philosophy of Technology, pp. 3–25.
© *1989 Kluwer Academic Publishers.*

between the bipolar factions who ('unintentionally') precipitated World War I. Where one nuclear power reacts to a signal of a possible enemy missile launch by putting its nuclear forces on alert when that enemy in fact did not launch, then the enemy will construe this alert as a threat to deploy those forces, i.e., missiles. Notice how a military analyst describes this upward cycle as a rapidly escalating divergence between *intent* and *effect*:

> Tight coupling of forces increases, information begins to inundate headquarters, and human, preprogrammed computer, and organizational responses are invoked. Each response, whether it arises from a human operator or a computer, is intended to meet some narrow precautionary objective, but the overall effect of both Soviet and American actions might be to aggravate the crisis, forcing alert levels higher ... Beyond a certain level not only is a mutually reinforcing alert possible, but there would be major political consequences as well. The alert itself would undoubtedly contribute to the political tension because, fundamentally, decision makers would know how impossible it is to separate military moves, however precautionary, from political ones ..., and this very incomprehensibility would drive the mutual stimulation. In particular, if the Soviet Union, which inactively operates its nuclear forces, actually seemed to be going on alert, this would be a shock that would trigger hundreds of preprogrammed American responses.[5]

Yet, if we *know* of this discrepancy between our first-person descriptions of 'precautionary' mobilizing and the enemy's second-person descriptions of 'aggravating' behavior, then we ought to remove the objective conditions for the discrepancy insofar as they increase the likelihood of accidental launch. Whether or not the government decision procedures are heavily bureaucratized, cybernetic theory can only pinpoint the *fault* for the failure of information sharing in the bureaucratic structure itself. Does it not matter that the failure might (in fact, inevitably will) result in accidental war? As studies of Nazi genocide have shown,[6] it is one thing to say that 'the information stopped here' in a bureaucracy, but quite another to say that 'this bureaucrat was responsible' for that information systems failure. Militarily, this shift away from agency has been quite recent in U.S. nuclear bureaucratic structure; as late as 1982, Admiral Hyman Rickover stationed identifiable, accountable officers at every decision node of the nuclear launch sequence.

A second reservation regarding the cybernetic framework is that bureaucratic explanations are more tractable when bureaucratic structure is transparent. There is a known hierarchy in the DoD command-and-control structure. There have been documentable decisions by knowledgeable individuals regarding these crucial gaps in information. Thus, my inquiry is only indirectly about the nature of such self-deception of bureaucrats; presumably, as my eventual Aristotelian analysis holds, self-deception may be parsed as culpable ignorance of the motivational principles or the consequences of some action. Nor is my inquiry about an amoral 'systems' view of the bureaucratic nuclear structure. I am inquiring only about actual hardware and software structures that produce the objective dangers of accidental launch, and the public relations strategies that hide those dangers.

II

The Pentagon's self-admitted horror stories regarding wartime communications[7] have been referred back to their own censorship board. As my former congressman (Wylie, R-OH) abolished Playboy-in-Braille, which at least gave a faint outline of naked reality for the benefit of the blind, the moralists have turned away the strategically curious from the Library of Congress shelves. For this reason, the following data constitute a very conservative estimate of the current probability of accidental nuclear war, reflecting pre-censorship trends. For reasons to be explained, the last available NORAD data on routine Missile Display Conferences (MDCs) and Threat Assessment Conferences (TACs, full-scale alerts) are for 1983.[8] Moreover, studies using the 1977–1983 data obviously do not factor in the new Star Wars multiplier effect. Caspar Weinberger, whose four-day notice on the SDI announcement produced the reaction 'It's not a bomb, is it?,'[9] now says that decision time for activating the so-called 'first layer' of the new system will be two to three minutes. This is one-third to one-half the current total average decision time, the 'window of vulnerability' during which a (supposedly) retaliatory U.S. attack must be ordered, in current accidental launch studies. For reasons of political emphasis, I will cite cases allowing zero decision time.

One caveat regarding the following overview of these studies is in order: whether in its original MAD version, defending U.S. population centers, or in its recently-scaled-down version of mere missile or 'point' defense (i.e., consistent with the 'enhanced deterrence' or 'NUTS' concept), SDI bears enough phenomenological similarity to enemy perceptions of first-strike scenarios that it will multiply the possible occasions for possible false alarms,[10] and consequently will have a multiplier effect on risk assessment.[11]

The remainder of the section is a philosophical fragment of the computer model first developed by Brian Crissey, for a workshop at the December 1984 Nuclear Weapons Freeze Campaign meeting, refined after collaboration.[12] (Crissey, a former Pentagon computer modeler, uses some of the same variables noted in the classic mathematical study by Bernard Bereanu.[13])

TABLE I

Attacker		Soviet Union Missile		SLBM	
Assumptions					
FlightTimeLo	9	FlightTimeMe	10.5	FlightTimeHi	12
DetectTimeLo	1	DetectTimeMe	1.5	DetectTimeHi	2
EvalTimeLo	0.5	EvalTimeMe	1.25	EvalTimeHi	2
DualTimeLo	3	DualTimeMe	5	DualTimeHi	2
AssessTimeLo	0.5	AssessTimeMe	1.25	AssessTimeHi	7
LaunchTimeLo	3	LaunchTimeMe	4.5	LaunchTimeHi	2
MDCLenLo	1	MDCLenMe	1.5	MDCLenHi	6
TACLenLo	1	TACLenMe	1.5	TACLenHi	2
MACLenLo	1	MACLenMe	1.5	MACLenHi	2

Case	Pessimistic	Best Guess	Optimistic
Launch Point	0	0	0
First Detection	2	1.5	1
Evaluation Point	4	2.75	1.5
MDC Called	4.5	3.25	2
Second Detection	9	6.5	4
TAC Called	6.5	4.75	3
Assessment Point	11	7.75	4.5
MAC Called	8.5	6.25	4
Earliest Attack Point	10.5	7.75	5
UTOLT Point	3	6	9
EMP Point	7	7	7
Impact Point	9	10.5	12
Slack Decision Time	0	0	4
Launch under attack policy			
Decision Time	0	0	5
Informed Decision Time	0	0	4.5
Clean Decision Time	0	0	3
Clean & Informed Decision Time	0	0	2.5
Launch on warning policy			
Decision Time	1	4.5	8
Informed Decision Time	0	3.25	7.5
Clean Decision Time	5	5.5	6
Clean & Informed Decision Time	0	3.25	5.5

Parenthetical page numbers and tables in the following are keyed to this study.

The accidental-launch analysis proceeds genetically, by adding plausible ranges of values for identifiable phases of a presumed missile attack of any given sort, calculating how much human or computer decision time is available for any response to that attack. Here we will restrict ourselves to the case of Soviet submarine-launched ballistic missiles (SLBMs). Let us begin this study by analyzing some of the major variables in Table I.

(A methodological aside here: all simulations of missile-launch sequences have the same set of variables; their differences consist in the particular hardware characteristics of the enemy missile system in question.) The first relevant variable, from the initial processing of a NORAD signal from the Satellite Early Warning System (SEWS), *viz.* EvalTime, 'could take as long as two minutes in the worst case of multiple, confusing signals' (p. 2). At the end of this time, humans (now?) or machines (under a computer-launch policy) decide whether to call a Missile Display Conference (MDC); this is assumed in the model to take only 30 seconds, and is an historical average. This is called the 'MDC called' point, while the MDC itself is presumed to last from 1 to 2 minutes. Note, in Table II (=Table VI, p. 21), that the geometrically upward trend in MDCs suggests 295 for 1984, while a letter from NORAD to the Center for Defense Information (CDI) lists the number at 155.[14] That 'the [real] trend is steadily upwards by about 40% per year and might be explained by increased sensitivity of sensors, increased missile testing, or revised criteria,' is no longer completely true. There may have been a purely political decision to increase the threshold at which infrared signals – i.e., possible enemy missile fuel 'tails' – are to be counted as worthy of further analysis. No more incidents of geese causing alerts. But Crissey has communicated that real trends in enemy ballistics argue for *lowering* that threshold. Whereas these studies assume the current value of two to three minutes as 'the boost phase [when] the missiles are infra-red-visible and vulnerable' (hence incidentally showing also how quickly an SDI 'killer' ABM would need to strike it before its MIRVs are released) 'the boost phase may in the future last as short as 60 to 90 seconds.'[15] Moreover, in military terms, miniaturization of warheads, cruise missiles which fly under conventional ground radar, new depressed-trajectory launches, and anti-detection (or 'stealth') delivery systems all should revise that threshold downward. Thus the sudden drop in the official count of MDCs does not reflect military realities but only revised MDC criteria. We can count this as a case of governmental deception, however, only if we can independently confirm that projectible consequences of official policy are contrary to official claims. More on this in my conclusion.

TABLE II

Year	Routine	MDCs	TACs
1977	1567	43	0
1978	1009	70	2
1979	1544	78	2
1980	3815	149	2
1981	2851	186	0
1982	3716	218	0
1983	1479	255	0

Source: Brian Crissey, Michael Wallace, and Linn Sennott. 'Accidental Nuclear War: A Risk Assessment,' in *Journal of Peace Research* 23, 1 (1986) [Table VI, p. 21].

At this stage of the detection process we encounter the philosophically interesting phrase 'dual phenomenology.' Under the arguably current 'launch under attack' policy of the U.S., confirming data are required in order to move to the point determined by the Commander in Chief of NORAD as 'Second Detection.' After the first (SEWS) detection, a ground-based radar takes from '3 to 7 minutes, depending upon which sensors have picked up the launch and which trajectory the missile is following' (p. 14), a variable called 'DualTime.' If CINCNORAD decides that these radar data confirm the SEWS data, a Threat Assessment Conference (TAC) is called. This Threat Assessment Conference of senior personnel 'is presumed to last from 1 to 2 minutes (TACLen),' culminating in an 'Assessment Point' at which a 'high assessment' value to the dual phenomenology would call in the President to decide what to do about such a presumed attack. This point is a Missile Attack Conference (MAC), where essentially a decision is required whether to 'use or lose' our own missiles. Adding up the above variable times in the warning process, it is obvious that this 'Use Them or Lose Them' (UTOLT) point is historically diminishing. Even if our own launch time for retaliatory missiles is only three to six minutes (depending on which system of our triad we use), as the model assumes, current trends in missile technology are closing the time during which the MAC is enabled to use those missiles.

Unique to the current trends is the diminishing decision time required for such dual phenomenology. First, there is a point in time when the electromagnetic pulse from enemy warhead detonations (EMP Point) will interrupt the C^3 system with a 50,000 volt per meter charge within a 1500 mile radius of detonation (p. 11). Such EMP effects could occur, crippling the communications between the four command centers to which SEWS data are sent, either at *or before* an adequate decision time, under the current dual phenomenology policy. Hence the debate about the adequacy of the dual phenomenology system concerns whether to launch our own presumed retaliatory missiles before we are known (or merely confirmed) to be under attack.

As Table I (above) shows, this move to the so-called 'Launch On Warning' policy involves a tradeoff between two crucial values: reliability of the warning information from two sets of warning sensors versus increased decision time. I shall refer only to the last measure of the latter, 'Clean and Informed Decision Time,' the only parameter involving Presidential participation in decisions. Comparing its 'best guess' values between the Launch Under Attack and the Launch On Warning policies, this time increases, if we assume the Soviets are launching only an SLBM attack, from 0 to 3.25 minutes. As incredibly short as this retaliatory deliberation time may be, it is better than literally nothing, which is exactly the total 'clean' time our Pershing II affords the Soviets under either launching policy – as Crissey *et al.* calculate separately. In the Pershing case, and (as Table 1 shows) in the case of the Soviet SLBMs off our east coast, there is virtually no 'cool' decision time, *regardless of any future improvements in any conceivable automatic-launch software.* Even if there were, humans are known to act, under intense time pressure, less rationally than otherwise.[16]

Assuming a current Launch Under Attack policy, these values belie U.S. claims at Geneva that Soviet ICBMs (SS18s) are the 'most destabilizing' part of the arms-reduction equation. As Crissey *et al.* independently show in another table, we already have 15.25 clean minutes to deliberate under the dual phenomenology LUA policy, and increase that to only 20.25 minutes under LOW. The comparisons between the SLBM and ICBM cases also show two other military or strategic entailments. First, there is some reason to move from 0 to 3.25 'clean' SLBM decision minutes, i.e., to a Launch on Warning policy, in order to get beyond the mere MAD posture to the real possibility of retaliation advertised in our 'enhanced deterrence' posture. At zero minutes, need we say, there is no possible retaliation. Second, there is even more strategic reason to remove that SLBM threat by an ability to knock out the Soviet satellites that guide those SLBMs. In other words, the first defensive motivation carries a real risk of accidental LOW, while the second ASAT capability is a preemptive motivation that may seem not only offensive but phenomenologically indistinguishable from the defensive, from the viewpoint of our enemies at their radar screens. Even if our motivations are defensive, a LOW policy (or even capability) may seem offensive, thereby exacerbating the divergence between original intent and ultimate effect of that policy. Hence my opening complaints about the amorality of the cybernetic theory of corporate action.

In this larger context of running known or knowable risks, I want to reply to a criticism (stated by Thomas Grassi)[17] regarding the possible irrelevance of my cited SLBM risk data. The criticism holds that there is objective value in having a declaratory doctrine that 'neither confirms nor denies' a LOW policy, in DoD parlance; i.e., that refuses to renounce LOW publicly. That declaratory doctrine has real deterrent value, in preventing intentional attack by an enemy who is presumably too rational to risk U.S. retaliation. The operative U.S. policy could be different at this juncture of enemy SLBM engineering; in fact, knowing that the degree of SLBM inaccuracy (CEP, 'circular error probable') is very high, we would not be under pressure to launch on warning, because chances are overwhelming that our retaliatory ICBM capability would not be impaired by letting those SLBMs fall where they may. (It is apparently morally acceptable, on this objection, that they fall anywhere but on U.S. missile silos.)

There are both long- and short-term responses to this objection. In the long run, assuming that Soviet SLBM technology will reduce their CEP problem, the cited risk data become more relevant over technological development time. But even in the short run, these data force us to admit the motivational relevance of my initial cybernetic issue regarding the misreading of enemy intent. LOW and ASAT may look to the Soviets like pieces of the same puzzle: their radar readers must ask whether we are attempting (achieving?) a first strike (capability?). Their hermeneutic principle here has a history: it is because the Soviets declared a ban on ASAT tests in 1983 that this suspicion of U.S. first strike intent arises. The AuCoin/Dicks amendment (December 1985) was, *inter alia,* an attempt to bar similar U.S. tests, i.e., to make the unilateral Soviet move a bilateral one, and thereby to change their hermeneutic principle (above). If they read certain radar signals as an accidental rather than an intentional launch, we run less risk of their retaliating. Thus, only if we declare no first use can they read such signals by this preferred hermeneutic. Where the very ability represented by our ASAT tests might seem to us like mere defense against enemy SLBMs, from their viewpoint it now seems like offense against part of their perceived deterrent force. Strategists on both sides know that SLBMs are guided by those satellites; to lose them is to lose what the Soviets may perceive as a retaliatory ability. Historically, in fact, their SLBMs were a retaliatory response to our Pershing II, an attempt to put us in equivalent peril. So there are objective, risk-reducing values, not merely political ones in bilateralizing once unilateral moves.

TABLE III

Probabilities of Unintentional [sic] Nuclear War
Assuming Crisis-Induced Launch-On-Warning Policy

Assumptions					
Sides	2	Alarms/Yr.	335 [est.]	SFRACT	0.04
LowATIME	6	LowRTIME	2	RATE	0.0734
MedATIME	7	MedRTIME	3	CLENGTH	7
HiATIME	8	HiRTIME	4	YEAR	1985
FTIME	4	DTIME	2.5	Crises/Yr.	0.75

ATIME	RTIME	PROB	TMEDIAN	TMAX	ANWPROB
6	2	0.6412	15	64	28.1%
6	3	0.6951	14	59	30.0%
6	4	0.7351	13	55	31.5%
7	2	0.5134	18	79	23.2%
7	3	0.5796	16	70	25.8%
7	4	0.6303	15	65	27.7%
8	2	0.4111	23	99	19.0%
8	3	0.4832	20	84	22.0%
8	4	0.5404	17	75	24.3%
			AVG ANWPROB		25.7%
			Expected Endpoint		1990 [as of 12/85]

Given this fundamental tradeoff between the accuracy of 'attack' assessment and 'informed' executive decision time, then, Crissey elsewhere[18] compares the possibilities of accidental launch under various attack scenarios. Table III reflects plausible values (low, medium, high) of 'arrival' and 'resolution' times for the bipolar missile systems we have concentrated upon above, our Pershings and their SLBMs. Assuming that false alarms are truly random events, i.e., that they are distributed evenly over every possible type of presumed attack scenario, these times are then weighted to reflect those two systems' actual percentage of their respective arsenals, and this allows a final calculation of Accidental Nuclear War Probability (ANWPROB) for a crisis-induced Launch On Warning policy. The rate of false alarms will depend obviously also upon the number of sides (1 or 2) who use that policy. Although it may be safe to assume that the Soviets do not now have a LOW capability, both because of their more primitive computer technology and because they immediately admitted an inability to respond to our 6-8 minute Pershing II flight times,[19] this is not to say either that they would not develop or deploy such a system. With Pershing II in particular, Table III assumes that, eventually at least, they would move to LOW.

Notice that the calculated degree of annual accidental nuclear war probability (ANWPROB) assumes that the 'Serious Fraction' (SFRACT) of false alarms, those that are not immediately dismissed at the MDC level, is only four percent. Although these calculations are conservative in light of their authorship date, *vis-à-vis* the assumed upward trends in false alarms and downward trends in flight times, they are quite frightening. Two of these variables which are currently purely political in nature, but which could also be partial functions of any move to Launch on Warning itself, are Crisis Length (CLENGTH) and Crises/Year, valued respectively at 7 days and .75 per year, as historical averages of known military (NORAD) or political (superpower) crises. That is, as in the foregoing discussion of cybernetic theory, the misperceptions of enemy intent deriving from the lowered certainty of enemy attack assessments can themselves induce crises.

On the contrary, Crissey *et al.* also calculate the ANWPROB for a quite different set of launch policies. (Since it employs the same method, I will not cite it separately.) If Launch On Warning is renounced, and our Pershing IIs are pulled back unilaterally, the ANWPROB is only 8.5%. The political steps necessary to backing down from this literal precipice of accidental war, self-evident from these results, have now materialized with the INF treaty. Contrary to the above objection that there is an objective value in having a declaratory LOW option, however, both of these conditions are necessary to achieving that lower risk figure. There are real, not declaratory, decision time differences between the LUA and LOW policies, differences that dictate the relevant degree of risk for each launching policy. Insofar as rationality (not to mention morality) requires that an agent minimize risk, and our enemy can assume our sanity, they will eventually infer that any declaratory LOW policy cannot be a real policy; i.e., that the declaratory policy must be a bluff. In retrospect, although it is important not to downplay INF with the State Department's canonical line that 'It's only four percent' of superpower arsenals, it is equally important not to overplay INF with the claim that it has solved the accidental-launch problem. Both the 'hardware' (Pershing) and the policy problems must be resolved. More telling in this latter regard is Richard Perle's dramatic reversal from his halcyon days of rationalizing Pershing deployments. As intimated by his cryptic post-INF aside on NBC's 'Face the Nation' (December 1987), Pershing II pullbacks 'will reduce hair triggers' in Europe.

III

Returning to our original queries of intentionality and culpability, then, let us dissect a representative instance of the U.S. governmental reply. This will license a hermeneutic for analyzing official attitudes toward the known risk of LOW.

The Pentagon's recent 'nuclear winter' study claims that it would take a much higher threshold of nuclear detonations to produce the effects claimed by the original independent nuclear winter study. A spokesman from the National Center for Atmospheric Research has concluded that their 'more sophisticated model' shows what 'looks more like nuclear fall than nuclear winter. We aren't getting the extinction-of-humanity scenario.' But even this higher threshold of nuclear extinction, according to former ACDA chief Kenneth Adelman, simply 'underline[s] the immediate need for missiles whose accuracy would allow lower-yield warheads.'[20] Let us examine this technological solution to the current state of nuclear insecurity. Its advocates assume a direct proportion between missile yields and damaging collateral effects – 'Smaller bombs are safer' – while its critics argue that other experimental data and simulations[21] show that this takes the wrong sample as its data base. That is, advocates do not bother factoring in the very different environmental makeup of the troposphere, the biological life-world, where small warheads are supposed to inflict their damage. A smaller blast would carry fewer side effects 'safely' into the upper strata, away from immediate human consumption; it increases lower-level poisonous ozone while it decreases upper-level ultraviolet protective ozone content. A larger blast would create a larger vacuum for removing the worst immediate side effects. That is, there is an inverse proportion between bomb size and immediate lethal side effects. The 'interference effects' of a different layer of the atmosphere present what Ian Hacking calls a 'fortuitous universal,'[22] having different properties from the data base used in the universal 'smaller is safer' claim. It looks as if the 'safe bomb' advocates are culpably ignorant, in other words, neither knowing nor wanting to know the evidence against their position.

We could parse this ignorance, then, in terms of Alan Donagan's distinction[23] between responsibility *prima* and *secundum quid*. Either the agent or the moral or political system is the cause of the relevant ignorance. Were strategists indeed concerned about the possibility/probability of accidental launch – assuming that nuclear winter is not a desirable strategic consequence of policy – then this 'safe bomb' argument would defy rationality. It makes accidental launch more rather than less likely, in that because they seem safer to use they become easier to use. Here is our cybernetic perception-response mechanism again, but now in the first-person or Aristotelian case of self-deception. It will be difficult to escape individual responsibility, then, even if one is a member of a given political system of bureaucracy. (This cybernetic reading of the situation is compatible with the DoD denial of a preemptive first-strike *policy* of striking in crisis with 'small' Pershing or MX warheads: they may still invite attack.)

My opening problematic of (personal or political) deception can thus be transposed to the problem of culpable error. Given these results from the accidental launch studies, and the apparent willingness to deploy 'limited' means for what is mistakenly believed to be a means to achieving our nuclear deterrence strategy, there are two nested questions: (1) Is this a *culpable* error regarding the new belief in 'smaller' means to deterrent ends? And, if so, (2) Do we blame the agent(s) or the system for it? In light of the cybernetic mechanisms being invoked by this current smaller-bombs-are-safer behavior, our possible explanatory options are basically two. Either rationality is not a valid assumption regarding policy discussions, and (say) psychiatric concepts of motivation will have to do, or rationality is relevant but we cannot assume any *moral* concern. Some other concern, such as the attempt to grab the lead against the Soviets absolutely or at all costs (regardless of consequences), would have to direct our inquiry into DoD motivations. A merely instrumental rationality would alone be relevant, since no consequentially sane person would risk omnicide in the pursuit of a political goal. Sanity is psychiatrically defined as regard for consequences.

I propose therefore to offer an Aristotelian notion of 'cost' that avoids standard objections to utilitarian forms of consequentialism. Rather than claim that individual *actions* have positive or negative consequences, *qua* utilitarianism, it would be safer to say that policymakers' *values* (or Aristotelian preferences, *hairetai*), as dispositions to act one way rather than another, have consequences. The smaller-weapons-are-safer argument exhibits what could be called 'productivist' or technocratic values: the solution to any strategic policy problem is simply to produce a technological gadget. Humanistic values, as contrary dispositions, would see the relevant solution as bilateral negotiations, or the aforementioned withdrawal of Pershing missiles and/or rejection of a declaratory LOW policy; i.e., as anti-productivist or anti-technocratic policy options. Consider, in terms of this value contrast, the following transcription and commentary by Captain William Moore, in a recently declassified Strategic Air Command briefing in 1954:

In his memo [Moore] underlined heavily ... the remark of SAC General Old [who] 'stated that the exact manner in which SAC will fight the war *is known only to General LeMay and that he will decide this matter at the moment, depending on the existing conditions.*' [sic]
  Then [General Curtis E.] LeMay took questions himself.
  Q: 'How do SAC's plans fit in with the stated [sic] national policy that the U.S. will never strike the first blow?'
  LeMay: 'I have heard this thought stated many times and it sounds very fine. However, it is not in keeping with United States history. Just look back and note who started the Revolutionary War, the War of 1812, the Indian Wars, and the Spanish-American War. I want to make it clear that I am not advocating a preventive war; however, I believe that if the U.S. is pushed in the corner far enough we would not hesitate to strike first.'
  [Moore concluded that] 'SAC is, in effect, a sort of "elite corps" dominated by a forceful and dedicated commander, who has complete confidence in SAC's ability to crush Russia quickly by massive bombing attacks. No aspect of the morals or long-range effect of such attacks were discussed, and no questions on it were asked. ...'[24]

As Moore concludes here, I will simply indicate the historical U.S. strategic preference, or policy valued, as 'preemptive strike'; what these strategic discussions have ignored is morals or long-range effects. That nobody discussed or raised the question of those possible effects invites a non-utilitarian or Aristotelian concept of intent: although the knowledge of the consequences of some action may remain potential or unconscious for any particular agent, we may under certain conditions attribute responsibility for having actual knowledge of those consequences.[25] Aristotle defines culpable error as that wherein: 'the cause of one's ignorance lies in oneself,' as opposed to the case of a mere 'misadventure [*atuchema*] where the cause lies outside oneself' (*Nicomachean Ethics* 1135 b 18–19). We know, thanks to the above risk assessment studies, the approximate chances (*tuche*) of misadventure or simple accident. But trying not to know, or simply to ignore what one does know, are both morally culpable states of mind. General LeMay is neither crazier nor more culpable than his current counterparts on the NSC staff; he is simply offering perspective on the perennial values or policy dispositions of U.S. strategic personnel.

Let us turn, at this stage of the information/theory dialectic about strategic intent, to the legal definition of negligence based upon the above Aristotelian definition of culpable ignorance. The Model Penal Code defines 'negligently' as follows, under General Requirements of Culpability:

A person acts negligently with respect to a material element of an offense when he should be aware of a substantial and unjustifiable risk that the material element exists or will result from his conduct. The risk must be of such a nature and degree that the actor's failure to perceive it, considering the nature and purpose of his conduct and the circumstances known to him, involves a gross deviation from the standard of care that a reasonable person would observe in the actor's situation.[26]

This embedded dispositional assumption about a 'standard of care that a reasonable person would observe' is quite Aristotelian; it may be an agent's fault that some crucial knowledge remains potential rather than actual (see *Nicomachean Ethics* 1146b14–1147b19).[27] Moreover, 'the material element of an offense' is defined without reference to agent motive. It may be an agent's *purpose* to achieve this material element, known to include unjustifiable risk, or one may act *knowingly*, being practically certain of risk though purposely achieving a different material element or goal. Assuming that strategists know the same consequentialist data as these independent studies show, our strategic posture under a LOW policy is at least knowing in this sense.

Obviously, one may regard *any* non-negligible probability of ANW or (*ipso facto*) potential omnicide as an unjustifiable risk, and thus legally/morally culpable. That just shows how far the disarmament process has yet to go. But as a first rudimentary step, the crucial difference between the high (26+%) and the low (8.5%) ANW risks (or, legally, the 'material element' of the difference between those risks) is defined by the ANW studies as the Pershing deployments and a LOW launch policy, as cited above. It is, unsurprisingly, on precisely these two counts (as well as on MX, now undergoing deployment) that the DoD testimony and press releases have been systematically misleading outside analysts. We can tabulate three important instances as a clue to our initial question whether the Pentagon might have some culpable strategy for knowingly discounting relevant facts regarding accidental-war risks. Accordingly, these instances will be stated as culpability queries rather than as facts.

(C.a) Are Pershing II missiles in Europe destabilizing regarding enemy perceptions or misperceptions of our strategic intent, and *a fortiori* regarding their possibility of accidental launch?

Here is an interesting retrospective on a case in point. An offer by Italian prime minister Bettino Craxi, to declare a moratorium on U.S. Euromissile deployment, was ignored by the *New York Times*. As it was elsewhere reported: 'Amid fear that the key NATO [deployment] decision might begin to unravel as a result of Craxi's proposal, the State Department said that the U.S. "rejected the concept of a moratorium or pause" because it could "seriously hinder" negotiations with Moscow to limit or eliminate intermediate-range missiles in Europe.'[28] Of course, there were no such negotiations, since the deployments prompted a Soviet walkout. The *New York Times*, as virtual mouthpiece for DoD values, later effectively denied that Craxi had ever made the offer. The whole episode has been a classic case of discounting things which discredit the *Times*'s nuclear-NATO cheerleading, of selecting NATO statements opposing Craxi's concept of withdrawal. Craxi's original promise to bring up the moratorium offer at the next NATO meeting was mysteriously ignored or scotched from the agenda. Although we later saw titles like 'U.S. offers Soviets arms reduction plan,' that admit 'administration officials have been divided over whether the U.S. needs [sic] to have the Pershings in Europe,' this is with the hedge that they 'are waiting to see if Soviet negotiators follow up that suggestion [of limiting both sides' Euromissile deployments] and incorporate it into their proposals in the Geneva arms talks.'[29] As the word 'needs' was undoubtedly handed down from U.S. officials for this press release, it corroborates a common observation among peace/technology researchers that the military think tanks do all the same studies that are published independently; i.e., they have the same consequentialist data. Yet I must update this example somewhat, in light of the new political fact of the INF agreement.

(C.b) Is current MX deployment destabilizing regarding ANW?

Although MX was not factored into cited ANW risk assessment studies, even if Pershings are removed the similar warning-time MX characteristics would keep the overall bipolar ANW data at roughly the current level. With its 'sitting duck' basing system, for the first time in world history the U.S. has a first-strike-capable weapon which invites rather than preempts attack on our own soil, whether or not it is designed for first-strike use. In his important

'Background Paper on the Probability of a United States Launch-on-Warning Policy for Strategic Land-Based Missiles,' former missile engineer Robert Aldridge cites the following exchange between Senator James Sasser and then DoD Secretary Weinberger, which intimates that C.b is simply the current instance of the question:

(C.c) Does the U.S. have a LOW policy?

Sasser: Mr. Secretary, the only way to make the MX invulnerable ... would be to launch the missiles on warning; isn't that correct?

Weinberger: No sir. No it would not, to launch it after attack because the attack could not be effective enough to destroy those missiles.

S: In other words, you would wait until actual nuclear attack took place on the United States.

W: We are talking a lot of hypotheticals, and we are verging on the edge of material that ought to be covered in a closed session. But we do not have a first strike plan. We do not have any desire or any strategy that requires it. ...[30]

Aldridge shows how Weinberger 'was not completely truthful,' quoting other official statements only slightly less subtle than the quote above from LeMay. But he explains also how Weinberger 'dodged the issue. First strike takes place before the Soviets launch an attack ... [i.e.] in the pre-attack phase. [LOW] and launch-under-attack take place in the trans-attack period, assuming that the warning is valid.' Semantically, it is one thing to plan a first strike, but quite another to launch presumed 'retaliatory' missiles with a false alarm as one's presumption, hence the 'mere' effect of a first strike. Legally, it is one thing to 'purpose' and another to 'act recklessly' (or negligently) regarding a first strike. The intertwined conceptual/legal problem here is twofold. First, legally: even if U.S. missile-launch officers know the general probability of accidental launch, would their pushing the button on any particular occasion constitute 'knowing' or merely reckless or even negligent behavior? Second, conceptually: is it possible to arrange an informational hierarchy in a bureaucracy, like the DoD chain of command, such that the missile launch officers do not (ever, by prearrangement) know the risk of ANW?

The legal aspects of this issue have answered themselves. The DoD has certainly been attempting to deny the public that information, while Weinberger *et al.* are being sued for same under the Freedom of Information Act. Since the cited ANW studies, the DoD has changed, and *classified*, the criteria for counting false (NORAD) alarms.[31] There is now a sudden but unexplained drop in the official number of radar/satellite errors causing Missile Display Conferences (MDCs), from 255 in 1983 to 153 in 1984. So, if a Trojan Horse now has only 153 legs, do you still call it a horse? NORAD also claims that, meanwhile, there have been no high-level Threat Assessment Conferences. Aldridge explains, however, that 'these are highly secret and it is not unusual for the public to be kept uninformed if circumstances do not force the matter to be revealed.'[32]

The above conceptual aspects also carry some related material evidence. Aldridge correctly sees that the ambiguity of Weinberger's testimony leaves open just this question of who is authorized to push the button under a LOW policy:

> It seems simple to dismiss past [NORAD] errors which have been successfully resolved. But it is not such a comfortable feeling when we look ahead to increasingly complex systems which govern decisions under a [LOW] policy. It becomes downright torture when we realize that systems are getting so complex that human decision making cannot enter the real-time loop, except as a token button pusher, and decisions must be realistically construed as being pre-programmed on computers.[33]

Moreover, it is logistically impossible in the current hair-trigger environment to find, inform, and prompt the Commander-in-Chief during a two- or three-minute 'window of vulnerability,' as if to expect that a launching policy could be realistically decided then. Is there such a policy of either computer pre-programming or DoD pre-authorization? In either case it is unparadoxical to say that DoD intends ANW. That is what the real or operative definition of LOW boils down to under current military/strategic conditions.

This strategic quandary helps us to decode an otherwise extremely cryptic allusion of new Supreme Court appointee Antonin Scalia. Addressing the media's role in guarding against 'an arbitrary executive,' he seems to pronounce the Reagan administration's position on precisely this matter of pre-authorization:

> The defects [of the Freedom of Information Act] cannot be cured as long as we are dominated by the obsession that gave them birth – that the first line of defense against an arbitrary executive is do-it-yourself oversight by the public, and its surrogate, the press. ... As a matter of fact, if you have a democratic process which has produced a classification system, the people have given it to the Secretary [of State? Defense?] and the President to keep information secret. It isn't that the press is less patriotic; it's that they haven't been designated by the people.[34]

Scalia's democratic principles have already been used to bury extensive evidence that 'the short endurance and rudimentary design of our control system makes controlled nuclear war impossible,' in the words of former Minuteman launch officer Bruce Blair.[35] After a Senate aide queried Blair on the 'DoD's sanitized version of your report' to the Legislation and National

Security Subcommittee,[36] Blair responded that deletions concerned Presidential launching options. In fact, if one reads all of the DoD '[censored]' tokens as either legally 'purposed' 'first strike' or even merely negligent 'launch on warning,' those substitutions would prove his claim that there is a *de facto* LOW policy. In a constitutionally unprecedented move, the congressional Office of Technology Assessment was barred from seeing its own authorized report, also by Blair. Described by an OTA advisor as 'the single most dangerous document I've ever seen,' 'The Pentagon quickly destroyed copies of the report ... and slapped a super-secret classification on it so that neither the author nor the ... Congress could see it.'[37]

IV

So much for the strategic facts. I will finally venture some conclusions about intentions, insofar as they relate to culpability and values as defined above. Blair offers, in our above sense, a productivist explanation for these strategic anti-communicative moves. The command system's 'effectiveness is hard to measure and doesn't provide large production contracts' to defense industries: it is passed over in budget requests 'because it cuts across the lines between the services *and strengthens civilian control*' (my emphasis). This motivational speculation may sound almost trivial by now; the intent of LOW is simply to keep the DoD in control, whatever the moral or strategic costs. Need I add that this is a basic principle of fascism? Or that Plato (*Republic* 576b ff.) regarded it as pardoxical even to analyze the tyrant, who was instrumentally rational but below ignorance concerning the right end or goal of the state?

Following his *Strategic Command and Control*,[38] Blair says that:

At present, we are operationally geared for launch on warning. ... Strategic organizations actually expect to receive retaliatory authorization within minutes after initial detection of [unverified] missile launches, [an] expectation so deeply engrained that the nuclear decision process has been reduced to a drill-like enactment of a prepared script ... whose purpose is to get a decision from the national command authorities before [presumed] incoming weapons arrive.

Moreover, every ritual presupposes a myth. Blair explains that:

Back in the 1960s, McNamara and others created this myth that we could design a command system that would allow us to ride out the maximum attack and respond deliberately. That was a myth ... but he made the case. And people, Congress, academia ... believed it, they thought that's what we had.[39]

Who are these national command authorities? Aldridge finally explains why the Pentagon takes itself as the legitimate presidential successor, as Scalia's above remark insinuates: 'predelegation of launch authority is mandatory for the quick response required' under LOW. Not the Presidential Succession Act but the Reorganization Act of 1958 is relevant to defining that authority, i.e., 'as establishing that the National Command Authority (NCA) for nuclear forces flows directly from the President to the secretary of defense.'[40] Alexander Haig's 'I am the vicar' remark, which created a brief media flap, was never understood. Just as former NATO Commander Haig mistook himself for the Secretary of Defense, and thus the legitimate successor under the presumed Reorganization Act, the DoD still regards it as true that any presidential mishap hands them the nuclear football.

In case Scalia's remark about 'defense against an arbitrary executive' does not speak for itself on this point, let me underscore an analogy between Blair's above example of the missile launch officers' 'drill-like enactment,' LeMay's 'I will decide this matter at the moment,' and Daniel Ellsberg's interpretation of Reverend Jones's suicide drills in Guyana:

> Jones called the practice sessions 'White Nights,' rehearsing his followers in the gestures of sacrificing their children and themselves, training them to react passively to his message. ...: 'Trust *me*. This time it's only a drill. I will decide ... when the time has come for us to meet together on the other side; the time for the cyanide.' That time finally came ..., just weeks before the NATO governments announced in December [1979] their decision, prepared in secret with no prior public discussion, the stationing of U.S.-controlled Pershing and cruise missiles on European territory.[41]

Under certain apocalyptic agency assumptions, namely that the DoD knows and advocates reliance upon the two worst-case factors regarding the high readings of the accidental launch data – *viz.* the LOW launching posture and fast-attack (MX) missiles – Ellsberg's Jonestown analog of U.S. launching officers is quite apt.[42] With LeMay's sense of history and policy,[43] Jones's corresponding sense of urgency, Scalia's sense and the DoD's practice of democracy, and finally the real risks cited by the ANW studies, all the terms of this analogy seem valid. It is possible to *intend*, in the legal sense of being culpably ignorant of known or knowable consequences, an accidental nuclear war.

In this connection, and as a conclusion, consider an incomplete analogy by Caspar Weinberger. Chiding the childishness of the peace movement, he appeals to the *Iliad* scene where the young Astyanax is frightened by the forbidding sight of an armored soldier approaching, unrecognizable with his helmet faceguard. The helmet suddenly removed, the son recognizes his heroic father Hector, and the whole family laughs, sharing 'a moment of peace.' By analogy, rational people are 'too wise to cry out at the sight of our protectors'[44] – *viz.* missiles. Weinberger failed to mention that Hector died soon thereafter in battle, and Troy was demolished, while in Euripides' account his wife Andromache heard that her young son was symbolically thrown to his death from Troy's highest tower. Weinberger refers to the 'enhanced deterrence' of our Homerically proportioned arsenal as 'our best hope.' His Homeric analogy forces us to compare, but not conflate, the domains of Aristotelian tragedy and ethics. The tragic hero is capable of repenting, when apprised of his error; although his tragic flaw has been translated (from *hate*) as 'blindness,' it is factual not moral ignorance. Only the knave is guilty of the latter, failing to repent even after knowing the facts.

University of San Diego

NOTES

* This is a technical version of my December 1987 paper, 'Preventing Accidental Nuclear War,' at the joint meeting of International Philosophers for the Prevention of Nuclear Omnicide (IPPNO) and the Concerned Philosophers for Peace, at the Eastern Division meeting of the American Philosophical Association. I am grateful to Douglas Lackey for feedback on an earlier version. Lackey's own 'Taking Risk Seriously', *Journal of Philosophy*, 83 (Nov. 1987): 633–40, was read too late to facilitate parts of my 'culpability' thesis.

[1] Alfred Mele, *Irrationality* (New York: Oxford University Press, 1987), chapters 8 and 9.

[2] Quoted by Flora Lewis, *New York Times,* December 3, 1985.

[3] I say 'accidental' rather than 'unintended' since, at least on the analysis of personal culpability to be endorsed, it is nonparadoxical to claim that an agent can *intend* to achieve the conditions which he *knows* (or, if negligent, ought to know) might result in an accident.

[4] John Steinbruner, *The Cybernetic Theory of Decision: New Dimensions of Political Analysis* (Princeton: Princeton University Press, 1974). His more recent 'Launch under Attack,' *Scientific American* 250 (January 1984): 37–47, might be read also as a relevant instance of the cybernetic theory, but will be modified as culpability requirements dictate. For a definitive statement of the amorality of the cybernetic or systems view, see the opening remark in Richard Ned Lebow's 'Loss of Control' in *Nuclear Crisis Management: A Dangerous Illusion* (Ithaca: Cornell University Press, 1987): 'I describe several different ways in which [loss of strategic control] could lead to unintended nuclear war between the superpowers. I describe the likely cause of each of these paths to war and show the extent to which they are structural attributes of the superpowers' alert and response systems. A high level of risk is, I shall show, inherent in strategic force generation' (p. 75). Nowhere has Lebow identified the

question of responsibility for such risks.

[5] Paul Bracken, *The Command and Control of Nuclear Forces* (New Haven: Yale University Press, 1983), pp. 65–66. See also his 'Accidental Nuclear War,' in G. Allison, A. Carnesale, and J. Nye, eds., *Hawks, Doves, and Owls* (New York: Norton, 1985).

[6] See H. Friedlander and S. Milton, *The Holocaust: Ideology, Bureaucracy and Genocide* (Millwood, NJ: Kraus International, 1980); Fred Weinstein, *The Dynamics of Nazism: Leadership, Ideology, and the Holocaust* (New York: Academic Press, 1984).

[7] Daniel Ford, *The Button* (New York: Simon and Schuster, 1985); excerpted in *The New Yorker*, April 1 and 8, 1985.

[8] D.W. Kindschi, Lt. Col. (NORAD), Aerospace Command Center, USAF, Public Relations Department. Letter to Gary Houser, 1983.

[9] See Frank Greve, 'Top Advisors Were in Dark on SDI Plan,' *Charlotte Observer*, November 17, 1985.

[10] See Kurt Gottfried and Richard N. LeBow, 'Anti-Satellite Weapons: Weighing the Risks,' in F.A. Long, D. Hafner, and J. Boutwell, eds., *Weapons in Space* (New York: Norton, 1986).

[11] Robert Aldridge, Dean Babst, and Linn Sennott, eds., *The Nuclear Time Bomb: Assessing Accidental Nuclear War Dangers through the Use of Analytical Models* (Dundas, Ontario: Peace Research Institute, 1986). See also Clifford Johnson, Dean Babst, Robert Aldridge, and David Krieger, 'Computer-in-Chief,' Global Security Study No. 2, Nuclear Age Peace Foundation (Santa Barbara, CA).

[12] Brian Crissey, Michael Wallace, and Linn Sennott, 'Accidental Nuclear War: A Risk Assessment,' *Journal of Peace Research*, 23 (1986): 9–27.

[13] Bernard Bereanu, 'Self-Activation of the World Nuclear Weapons System,' *Journal of Peace Research*, 20 (1983): 49–57. See also Alan Borning, 'Computer System Reliability and Nuclear War,' in *Communications of the Association for Computing Machinery*, February 1987; and Barbara Marsh, 'The Probability of Accidental Nuclear War: A Graphical Model of the Ballistic Early Warning System,' unpublished M.S. thesis (Monterey, CA: Naval Postgraduate School, 1985).

[14] David Morrison, letter from Lt. Col. Charles Wood, Acting Director of Public Affairs, Headquarters Space Command (NORAD), as reported in *The Accidental Nuclear War Prevention Newsletter* 1, 3 (September 1985). Available from Nuclear Age Peace Foundation, 1187 Coast Village Rd., Suite 123, Santa Barbara, CA 93108.

[15] Personal correspondence, January 1986.

[16] William Arkin and Peter Pringle, 'C^3I: Command Post for Armageddon,' *The Nation* 236 (April 9, 1983): 434–438. See also Morris Bradley, 'Psychological Processes That Make Accidental Nuclear War More Probable,' Technical Report No. 8, Nuclear Age Peace Foundation (see note 14, above).

[17] At APA convention (see note * above).

[18] From Crissey, chapter 2 of *The Nuclear Time Bomb* (Dundas, Ontario: Peace Research Institute, 1986).

[19] Alexei Aleksandrov, interview in the *Christian Science Monitor*, June 3, 1982.

[20] See *Newsweek*, March 31, 1986, p. 65. See Kenneth Adelman, 'Proposed Intermediate Missile Accord Doesn't Do Everything,' in *San Diego Union*, July 5, 1987.

[21] Carl Sagan and colleagues update these data on inverse proportionality in Lester Grinspoon, ed., *The Long Darkness: Moral and Psychological Analyses of Nuclear Winter* (New Haven: Yale University Press, 1986). See also Z. Kripke, 'The Health Effects of Underground Nuclear Testing,' San Diego *PSR Newsletter*, August 1986.

[22] 'Culpable Ignorance of Interference Effects,' in D. MacLean, ed., *Values at Risk* (Totowa, NJ: Rowman and Allenheld, 1986), pp. 132–152.

[23] 'Consistency in Rationalist Moral Systems,' *The Journal of Philosophy*, 81 (June 1984): 291–309.

[24] Quoted in David Rosenberg, 'A Smoking Radiating Ruin at the End of Two Hours: Documents on American Plans for Nuclear War with the Soviet Union, 1954–55,' *International Security*, Winter 1981.

[25] I discuss the logic of such cases in 'Moral Dilemmas, Deliberation, and Choice,' *The Journal of Philosophy*, 82 (March 1985), esp. note 15 and Section II.

[26] *Model Penal Code and Commentaries (Official Draft and Revised Comments)* (Philadelphia, PA: America Law Institute, 1985), § 2.02, p. 226.

[27] Earlier versions of the Crissey *et al.* studies (see notes 11–13, above) used the phrase 'unintentional nuclear war,' a title now amended to 'accidental.' Thus, with regard to the case at hand, it would be self-contradictory to say that one directly *intends* an unintentional war, but only paradoxical to say that one intends a strategy whose actual (probabilistic) effect is, legally, an 'unreasonable risk,' viz. of ANW. One may be held accountable for knowing that consequential entailment, even if one were 'actually' ignorant of it. On the relation of this point to Aristotle's theory of action, see Alfred Mele (note 1, above).

[28] *Los Angeles Times*, May 8, 1984. This episode is where my earlier study, 'The Representation and Resoluton of the Nuclear Conflict' (*Philosophy and Social Criticism*, 10 [Winter 1984]) concluded, and essentially where the present study resumes.

[29] *New York Times*, September 21, 1986, p. A1. If anyone is a real student of these media misrepresentations, I can furnish quotations from three months of *New York Times* microfilms on this Craxi incident.

[30] Robert Aldridge, 'Background Paper on the Probability of a United States Launch-on-Warning Policy for Strategic Land-Based Missiles' (Santa Clara, CA: Pacific Life Research Center, May 1986).

[31] In reply to a Freedom of Information (FOI) request from David Morrison of the Center for Defense Information, as reported in the *Accidental War Prevention Newsletter* 1,3 (September 1985), p. 4. See also note 14. See also Bill Moyers, 'The Constitution vs. 1987? Problems with the National Security State' (interviews with Bruce Blair and constitutional lawyer Edwin B. Firmage; seventh of PBS series, 'In Search of the Constitution,' 1987); and Clifford Johnson, 'The Constitution vs. the Arms Race,' *Computer Professionals for Social Responsibility Newsletter* (Spring 1987). The latter describes Johnson's lawsuit against Weinberger *et al.* over the DoD's unconstitutional usurpation of presidential authority for having a launch-on-warning capability; the suit has so far been dismissed as 'merely political' by a federal district judge. The case is also recounted by Edward Lempinen, *Student Lawyer*, May 1987.

[32] Aldridge (note 30, above), p. 18.

[33] Aldridge, p. 17.

[34] Quoted by James Ridgeway, *Village Voice*, July 1, 1986, p. 46.
[35] *Strategic Command and Control: Redefining the Nuclear Threat* (Washington: Brookings Institution, 1985), p. 235.
[36] 'Command out of Control? Working Profile: Bruce Blair,' *Nuclear Times* (May/June 1985), p. 235. In the report itself, Blair also notes that LOW is an official option, programmed into the war-fighting Single Integrated Operational Plan (SIOP). See *Our Nation's Nuclear Warning System: Will It Work If We Need It?*, transcript of House hearings (September 26, 1985).
[37] *Ibid.*
[38] *Ibid.*
[39] *Ibid.*
[40] Aldridge (note 30, above), p. 8.
[41] 'Call to Mutiny,' in E. P. Thompson and Dan Smith, eds., *Protest and Survive* (New York: Monthly Review Press, 1981), p. xvii.
[42] Unless, of course, the missile launching officers have been sensitized to the *angst* of pushing the button. Colonel Mal Wakin averred in a public debate with Richard Wasserstrom ('War, Morality, and Public Nuclear Policy,' University of Dayton, October 1983) that by teaching ethics to those officers the military would be humanized.
[43] See LeMay's recent claim that civilian control of atomic bombs 'hampered' the effectiveness of SAC at its inception; see also his claim that he did have secret access and is 'saying that [he] could have started the [sic] war' in the early 1950s – i.e., before the present command-and-control structure began to authorize a DoD origin to the 'Permissive Action Links' (PALS). The source is *International Security* (Spring 1988), and is quoted in 'Nuclear Attack Reportedly Would Have Taken Week,' *San Diego Union,* March 24, 1988 (A24).
[44] 'A Rational Approach to Nuclear Disarmament,' in James Sterba, ed., *The Ethics of War and Nuclear Deterrence* (Belmont, CA: Wadsworth, 1985).

DAVID BELLA

CATASTROPHIC POSSIBILITIES OF SPACE-BASED DEFENSE[1]

INTRODUCTION

Ballistic missile defense will require the rapid release of defensive weapons. This paper considers the consequences of mistakes if both the United States and Soviet Union deploy such defensive systems. Each system would pose a threat to the other. There is a danger that a mistake or event could trigger rapidly expanding exchanges between the opposing defensive systems. Through such interactions, defensive systems could self-activate and amplify relatively minor events to catastrophic levels. Assuring against such catastrophic amplifications while also assuring the effectiveness and survivability of such defensive systems is a necessary requirement that has not been adequately addressed.

A large and controversial research and development effort, the Strategic Defense Initiative (SDI), is investigating the feasibility of deploying a ballistic missile defense (BMD) system, referred to by some as 'Star Wars.' Space Based BMD systems proposed under SDI[1, 2] will be considered here. Such a BMD system could employ a variety of advanced weapons including satellite-based lasers, particle beams, and electronic rail guns. The proposed multilayered BMD system would intercept nuclear warheads throughout their trajectory. Of particular importance is the destruction of ballistic missiles in their boost phase [3–5].

The boost phase of intercontinental ballistic missiles lasts several minutes after launch. Fast burn boosters could reduce this time to less than a minute [6]. Thus, intercepting missiles in the boost phase requires extremely rapid responses from a BMD system. The advantages of boost-phase intercept, however, are high because each booster may carry 10 or more nuclear warheads. Boost-phase intercept is only one of the requirements that demand a highly responsive BMD system that can carry out extremely complex operations within very short intervals of time. This paper will examine the consequences of deploying such a rapid-response weapons system.

A primary reference for this paper will be the 'Fletcher Report' [7], and in particular, Volume V entitled, 'Battle Management, Communications and Data Processing'[8]. The 'Fletcher Panel,' consisting of approximately 50

Paul T. Durbin (ed.), Philosophy of Technology, pp. 27–40.
© 1989 Kluwer Academic Publishers.

prominent scientists and engineers, came out strongly in favor of BMD. Their report, most of which is still classified, is frequently cited by proponents of BMD as an authoritative source.

RAPID RELEASE OF WEAPONS

It is apparent from a variety of sources and the technical requirements of a BMD system that the release (firing) of defensive weapons will be highly automatic [6, 8, 9]. Under the 'Fletcher Study,' the panel on 'Battle Management Communications and Data Processing' stated that *'some degree of automation in the decision to commit weapons is inevitable if a ballistic missile defense system is to be at all credible'* (emphasis is given by the panel). The panel concluded: 'The battle management system must provide for a high degree of automation to support the accomplishment of the weapons release function'[8].

Greater reliance may also be placed on more rapid release of offensive weapons, particularly during times of high tension [10, 11]. In Senate hearings, George Keyworth, then Science Advisor to President Reagan and a strong advocate of BMD, testified: 'We may well be required to accept a posture of launch under attack – 20 to 30 minutes, even less – in which we would have to make that critical decision to launch nuclear weapons'[9]. 'Critical decision time' available after an enemy launch is considerably less than the flight time of 20 to 30 minutes and could approach zero depending upon the nature of the attack [11, 12].

Assurances that 'humans will remain within the loop' do not remove the problems or difficulties associated with rapid release. A BMD system at a high state of alert is a highly complex and responsive system. In any future space-based military engagement, a delay of even seconds could yield significant military advantages to the opponent. Command authorities in the future would be under tremendous pressures for rapid release capabilities and a greater reliance upon automation should be expected [8].

One should consider such rapid weapons release mechanisms as essentially automatic even if there are buttons for humans to push. After all, the presumed purpose in keeping a human in the loop is to allow for meaningful real time assessment reflection, dialogue, responsibility, and common sense for these most crucial decisions. Many technical experts presume that a few minutes or even seconds is sufficient time for human intervention. This presumption may well constitute one of the most significant gaps between technical expertise and informed public opinion. BMD as now proposed does not accommodate such real time involvements. The consequences of mistakes in such a rapid response system must be seriously examined.

MISTAKES

Any system and its mistakes can be viewed from different levels of organization. One can view the components of a system or one can view the system constituted by the relationships between such components. Behaviors and problems at one level may not be apparent at another level. Solutions found reasonable at one level may be found to be unreasonable at another. A proposed BMD system will be viewed herein from three levels: component, system, and meta-system.

From a component level view, one focuses attention upon the component devices: weapons, sensors, and computers. Most proponents of BMD take such a component view. Their arguments are dominated by discussions of such devices and their performance. From a system view, on the other hand, one focuses attention on the relationships among such components [13]. Some of the strongest criticisms of BMD have been concerned with the software that will be needed to link the components into a functioning system [14].

Proponents of BMD recognize that mistakes will occur at both the component and the system level. Research is being directed to reduce the level of mistakes while making the BMD system fault tolerant so that it may function despite mistakes [1, 2]. As an example, a panel of computer experts recommended a 'decentralized' or 'distributed' system architecture that would allow a BMD system to function reliably despite the presence of some error [15].

The potential mistakes that have dominated the BMD debate are those mistakes that would prevent the destruction of enemy warheads before they reach friendly territory. Consider another category of mistakes, those that might occur prior to hostilities. Proponents of BMD acknowledge that such mistakes could occur. They reason, however, that the consequences of a peacetime BMD failure are relatively insignificant. Moreover, they claim that defensive systems can reduce the terrible consequences of an accidental launch of a nuclear armed missile [16]. Defensive weapons, they emphasize, are designed to attack weapons, not people. If they misfire during peacetime, they are not likely to hit anything of great importance. This rationale may be paraphrased as follows:

It does not matter greatly if an occasional peacetime misfire occurs within a BMD system because the consequences of such mistakes are relatively insignificant. Defensive weapons are not weapons of mass destruction; they are designed to attack weapons and not people. In comparison to the destruction caused by even one offensive nuclear warhead, the misfire of a defensive weapon is insignificant. A shift toward a defensive system involves a tremendous reduction in the consequences of mistakes.

The development of a BMD system as now proposed is based upon this rationale. This rationale is clearly central to the positions of George A. Keyworth II, then Science Advisor to President Reagan, and Richard N. Perle, Assistant Secretary of Defense for International Security Policy, given in Senate testimony in response to questions by Senators Joseph R. Biden and Paul E. Tsongas [9]. Defensive misfire is either ignored by proponents of BMD or briefly addressed and then dismissed [3, 5, 17]. From a component or system view, this rationale does indeed appear to be reasonable. BMD weapons are designed to attack weapons and not people. But, does this reasoning hold up when BMD is examined at a broader level of organization?

A META-SYSTEM LEVEL ASSESSMENT OF MISTAKES

The meta-system is defined as the global complex of interacting military systems of opposing sides. At the meta-system level, system boundaries encompass the strategic systems of both the U.S. and U.S.S.R. The strategic systems of each nation are thus subsystems of a global strategic meta-system and the interactions between these subsystems are of primary concern [18]. Proponents of BMD (notably President Reagan) reason that the world would be safer if both the U.S. and U.S.S.R. deployed a BMD system. Such reasoning reflects a component and system level analysis. From a meta-system view, the interaction between such BMD subsystems must be examined.

The technology now being studied for BMD, particularly for boost-phase and mid-course intercept (rockets, lasers, particle beams, sensors, and computers), could be well suited to attack enemy BMD forces, particularly their satellite components [6, 19, 20]. The technological task of destroying satellites, which follow predictable orbits, is likely to be simpler than the task of boost-phase intercept. If we assume that both sides place a high value upon BMD technology, then each will perceive the BMD system of the other to be a serious threat that demands protective actions including weapons for active defense.

In any BMD battle, time would be of the utmost importance. With satellite sensors, high speed computers, and weapons that strike at or near the speed of light, a delay of even seconds could be catastrophic. Thus, in time of crisis and high states of alert, the sensor-computer-weapon systems of both sides would be highly sensitive to any actions of the opposing side. The two BMD subsystems would be tightly coupled. Now consider the consequences of a mistake. Such a mistake will be called an initiated event. In a high state of alert, there is a significant probability that somewhere within the vast strategic system an initiating event would trigger some sort of reaction from within one or both of the BMD subsystems. This reaction could, in turn, promote a response from the opposing side. Within seconds, an amplification of the initiating event could occur through a series of responses to responses to responses. Paul Bracken examines such amplifications within tightly coupled command and control systems [10]. Charles Perrow examines the behaviors of complex and tightly coupled systems and provides examples of such instabilities [21]. Such amplification would occur when the response to a perceived threat is interpreted by the other side as a greater threat which in turn produces a response. This new response in turn triggers a still greater response from the other side. The process continues, producing a systemic and largely automatic amplification that rapidly becomes independent of the initiating event.

In a state of crisis and high alert, the stage is set for such an amplification because each side readies its own sensor-computer-weapon system to be highly responsive to the threat of the other. Reliance upon automatic systems is high. The systems of both sides are programmed to take threats seriously and respond with equal or greater force to any hostile action from the opposing forces. These are the very conditions required for amplification within the BMD meta-system.

Given the speed and precision of BMD weapon systems and the threat they pose to each other, both sides would have strong incentives to go to higher alerts thus increasing the amplification potential of the meta-system. Either side could reduce the amplification potential of the meta-system but only by placing itself at greater risk to the threats of the other. Both sides would have deployed a technological system for which a delay of even seconds could be militarily catastrophic. Thus, each side would have strong incentives to decrease its own response times by going to high states of alert. But in going to higher alerts, both would face a greater risk that an initiating event, unforeseen by either side, could trigger an amplification within the meta-military system to catastrophic levels.

A global BMD system in a high state of alert is a system poised on a bifurcation point, capable of flipping into a condition of catastrophic consequences. The fundamental problem does not lie in the initiating event. The problem lies in a global strategic system that by its very nature has the potential for amplifying any such event. This amplification potential arises from those technologies that make a BMD system possible. Highly sensitive sensors, high speed computers, lasers, and particle beams that fire at or near the speed of light, redundant communication links, and a distributed system architecture that responds quickly without system-wide coordination; all these technological advances on both sides set the stage for the global amplification of an initiating event unforeseen by either side. The fundamental problem is not hidden in some unforeseen software error or some isolated computer malfunction; it lies in the operational requirements of each opposing BMD system.

THRESHOLDS FOR RAPID WEAPONS RELEASE

While the open literature contains very little on the rapid release of strategic weapons, a general approach can be identified from congressional testimony [9], the 'Fletcher Report' [8], and the demands of the technology now being developed. Command authority, with the President as Commander-in-Chief, would define threat levels, contingencies, and conditions that would serve as thresholds for defensive actions including the release of weapons. Any person 'in the loop' would be embedded within a complex technological system and highly dependent upon this system for information and assessment. During high states of alert, response times for this highly automated system would be extremely short. Thus, thresholds would need to be programmed into the computers and into any person in the loop through rules of engagement, procedures, etc. In theory, weapons would be released only when appropriate thresholds were exceeded.

A threshold thus serves to prevent the release of weapons below some threat level while assuring the release of weapons above some threat level. Different thresholds can be employed for different weapons. Thresholds can be kept high during quiet times and lowered by command authority when the possibility of attack appears to increase. Thresholds for rapid release of weapons would be based upon such conditions as the 'number of boosters in track,' 'whether defense resources are under attack' and the 'surviving status of defense resources' [8].

Under a section entitled 'Requirements for Automation,' the 'Fletcher Study' describes this threshold approach as follows:

It seems to the Panel that the battle management system must provide for –

An ability for the command authority to define thresholds or contingencies within which release of weapons is delegated to the automated systems. Examples are release nuclear weapons for defense of own resources, release hit-to-kill weapons if more than ten boosters are in track, and release all nuclear weapons if more than 100 boosters are in track ([8], pp. 25–26).

One should note that the release of nuclear weapons is twice included as an example.

Within a highly automated system, weapons are, in fact, released in response to sets (patterns) of signals and computer outputs from within a vast technological system. This is true even when human assessment is contained within the decision loop. The job of software engineers is to develop elaborate codes (computer programs) so that complex sets of signals can be rapidly processed to provide credible indicators of the threat levels, conditions, and contingencies defined by command authorities. This task is complicated by the fact that the first step by an adversary in any attack could involve disrupting and confusing the system that produces such signals and outputs.

By disrupting the command and control system, an adversary would hope to cause confusion so as to delay an effective counterattack. Given a high level of automation, the first phase of an enemy attack might seek to produce incoherence (inconsistency or incompleteness) in the signals and outputs within our own system so as to delay the release of our weapons. With BMD deployment on both sides, a delay of even seconds could be decisive.

The possibility of such an attack must be incorporated in the design and operation of a BMD system. One approach is to develop software that is capable of rapidly identifying such an attack so that responses can be made without costly delay. This is, however, an extremely difficult task because the sets of possible incoherent signals and outputs are extremely large. It is quite difficult to define in advance rules to interpret such signals and outputs so that such an attack could be identified and not confused with disruptions caused by space debris, equipment malfunctions, program errors, high tech harassment, sabotage and terrorism, and misguided but brilliant and well-equipped 'hackers.'

The point of this discussion is that when command authorities define some threat level (condition, contingency, level of attack) for which weapons release is authorized, they are in fact defining a technological task of immense complexity. Moreover, this difficult technological task would be never ending. A BMD system would be under continuous development and would need to respond to ever changing threats [15]. Margins of safety would be required. Safety margins to prevent the false withholding of weapons would result in lower thresholds. Safety margins to prevent the false release of weapons would result in higher thresholds. How would such thresholds be employed in a strategic weapons system? Answers await further research and disclosure. Nevertheless, a set of general assumptions reflected in the literature is outlined in Table I.

TABLE I
General Assumptions Concerning the Release of Strategic Weapons

1. Relatively high thresholds will be defined for the release of offensive nuclear weapons to assure against the catastrophic consequences of mistaken release.
2. Relatively low thresholds will be defined for defensive weapons to assure rapid release during a missile launch or an attack on the BMD system; such lower thresholds are acceptable because the consequences for the inadvertent release of defensive weapons are low.
3. Lower thresholds will be employed during times of crisis and high states of alert.

The general approach outlined in Table I seeks to sustain a highly responsive and relatively low risk defensive system while at the same time maintaining a high level of caution (high thresholds) with respect to the release of the more destructive offensive nuclear weapons. These are worthy goals. When viewed from a meta-system level, however, this general approach could be catastrophic.

A lower threshold for the release of defensive weapons increases the probability that a relatively minor event could be amplified through the interactions between BMD systems (much like the feedback in a public address system). The amplification could quickly reach the level where either one or both sides had evidence of 'unequivocal attacks against several defense resources.' The 'Fletcher Panel' cites such evidence as 'highly cautious criteria' to govern automatic release procedures during normal peacetime operation. 'During periods of tension,' the panel continues, 'lower thresholds might be set' [8]. Amplified responses would also threaten and disrupt command, control, communication, and intelligence systems which could, in turn, prompt the use of offensive nuclear weapons. Pressure to respond quickly would be enormous. The danger is that the amplifications of threats, triggered by an initiating event not intended by either side, would increase apparent threat levels above the release requirements or thresholds of some offensive nuclear weapons. The assumptions of Table I could lead to global catastrophe as illustrated in the following scenario.

A SCENARIO

Assume that both the U.S. and U.S.S.R. have deployed spaced-based BMD systems. Both have 'distributed' system architectures within which essential functions that must be performed automatically and rapidly need only be coordinated over a small area. Such functions are delegated to small battle groups consisting of several sensors and weapon platforms in close proximity. Higher level battle management is updated every several seconds with condensed information from local battle groups. Such a distributed or decentralized system architecture has been recommended by a Strategic Defense Initiative Organization (SDIO) panel called the Eastport Study Group [15]. The Eastport Study Group report [15] did not consider attacks against the defense system. Neither did they consider BMD interactions. Such limitations render their recommendations premature at best. General Abrahamson, SDIO director, has been quoted as saying: 'The SDIO endorses the spirit and content of the report of the Eastport Study Group. It is found to be in harmony with the needs of the SDI program and its rapid implementation shall be pursued throughout the R&D effort' [22]. Thus, the setting for the following scenario is not inconsistent with SDI efforts.

In a time of increased tension, procedures for such steps as maneuver and activation of sensors and weapons are implemented on both sides so as to permit rapid responses to threatening actions by the opponent. Thresholds for the release of defensive weapons are set to allow small battle groups to respond rapidly to attacks with a minimum of system-wide coordination. U.S. space policy states that 'purposeful interference with space systems shall be viewed as an infringement upon sovereign rights' [23]. The highest thresholds are set for offensive nuclear weapons with the lowest offensive threshold for nuclear weapons based upon conclusive evidence that the entire defensive system is under full-scale attack (see Table I).

Given the threats posed by the enemy BMD systems, each side sees real advantages to the above steps. The complexity for battle management software is reduced by reducing the need for system-wide coordination. Responses to an attack can be more reliable and rapid because delays required for system-wide coordination are reduced. Prompt action by local battle groups makes it difficult for an enemy to force a hole in one's own defensive shield. Battle groups can respond despite damage elsewhere in the system. Software errors have only localized effects. Finally, the high thresholds set for offensive weapons reassure both sides that the steps they have taken are prudent.

Somewhere within this global system, through mistakes and unanticipated events, the set of signals within the computers of a single battle group (it makes no difference which side) prompts some sort of action. This action, in conjunction with other unrelated events, such as a misperception by a person in the loop, is interpreted as exceeding some threshold within only a small portion of the global system, perhaps only a single battle group. Some reaction occurs. This reaction exceeds thresholds within additional portions of the opposing system. They react. The severity and scale of the reactions rapidly expand. Reactions include maneuver, repositioning, activation of sensors and weapons, and then, somewhere within the global system, 'defensive shootback.' The reactions include further shootback.

Central battle management cannot keep up with the expanding reactions. The humans in the loop are overwhelmed with incoming data. Responses on both sides become more automatic as indications of attack rapidly accumulate. Both sides are operating under conditions that were impossible to test realistically. On both sides, the perception, 'we are under attack,' promotes actions. Somewhere, it could be either side, offensive nuclear weapons are fired, perhaps only one. In response to an inquiry by Senator Sam Nunn, a Strategic Air Command (SAC) study concluded that, given the explosion of a single nuclear weapon at the height of an international crisis, the superpowers could respond in such a manner that global nuclear conflict became inevitable [24].

Either side could have reduced the probability of such a catastrophic outcome by setting higher defensive thresholds and imposing longer lag times for responses to permit more extensive real time assessments. However, such actions would have required accepting military disadvantages with no assurances that the enemy would make similar sacrifices.

CONCLUSION

Technological advances in sensors, high speed computations, communications, and space-based weaponry are transforming the global meta-system so that, in a high state of alert, the technological meta-system would have the capacity to unintentionally amplify relatively minor events to global catastrophe. Even if BMD systems are not deployed, this dangerous global transformation could continue as technological advancements make each side more dependent upon its own space-based assets and more threatened by its opponents. In brief, technological advancements are opening up new catastrophic possibilities.

When evaluating such catastrophic possibilities, one must keep in mind several minimum requirements. First, a trial-and-error approach is unacceptable. One cannot wait and see. Second, any proof that a catastrophic possibility can and will be avoided (or is impossible) must be overwhelming. The probabilities of catastrophic possibilities must be shown to be exceedingly small [25]. Third, the burden of proof must fall on those who are responsible for the technological actions that open up catastrophic possibilities. In this case, the burden of proof must be on SDIO. Fourth, any claim that the probability of catastrophic possibilities is sufficiently small must be clearly stated, openly defended, and able to withstand critical and independent review. SDIO has not met these requirements.

The research plans outlined by SDIO [1, 2] do not address the problem of global BMD amplifications. A 1987 review of SDI found that an analysis of two-sided BMD has not been satisfactorily done [26]. It has been reported that a recent and still classified independent review of SDI charged that insufficient analysis has been done on space-based threats [27]. Despite such insufficient analyses, a recent SDIO report states that battle management, command, control, and communications (BM/C^3) would support interaction between offensive and defensive systems [28]. SDIO claims that this interaction would allow the vital exchange of information between defensive and offensive forces 'during any situation to mutually enhance their performance.' In other words, SDIO has not adequately examined the technological instabilities of a global meta-system resulting from two-sided BMD deployment, and yet it has proposed to link offensive systems to future defensive systems. The catastrophic possibilities of such proposals cannot be found in the SDIO's assessments. (For two years, the author made numerous requests to SDIO and other federal agencies for reports that addressed the catastrophic possibilities discussed in this paper. The author reviewed all reports received and as of May 1988, was unable to find a single SDIO report that even discussed such catastrophic possibilities.)

Perhaps one should not be surprised by these deficiencies. The record of impact assessments in other fields provides evidence that organizations cannot be relied upon to assess the adverse consequences of their own activities [29]. These same organizations, however, are capable of bringing about large technological transformations. There is a fundamental problem here that goes beyond the feasibility of specific SDI proposals.

Technology is a social process that suffers from a fundamental imbalance of abilities. Technology's capacity to transform our world exceeds its capacity to assess the consequences of its own activities. A 1972 analysis of ecological impacts [30] arrived at similar conclusions. That is, technology's capacity to transform the world will exceed its capacity to foresee the consequences and thus, a strategy is needed to avoid catastrophic possibilities, preserve correctability, avoid large-scale irreversible change. Winner's assessment of technology [31] is relevant to the concerns expressed here. When the possible consequences are local, reversible, and non-catastrophic, a trial-and-error approach may be tolerated as a means of compensating for this imbalance. To a significant degree, technological development has come about as an adaptive response to unexpected consequences, deficiencies, and failures. However, as the power of technology expands, possible consequences of technological failures become global, irreversible, and catastrophic. The trial-and-error approach becomes intolerable. Given catastrophic possibilities, how might technology then adapt?

Technology could adapt by limiting its efforts to 'inherently safe' systems for which catastrophic failures are impossible. It could consider only 'fault tolerant' systems which are only capable of non-catastrophic failures despite unforeseen events, faults, and errors. Such approaches would place severe constraints on space-based defense [25]. However, the technological process could also adapt by *insulating* itself from adverse assessments, *dismissing* catastrophic possibilities, and *sustaining illusions* of its own capacity to predict and control events. To date, it appears that SDI is an example of this latter adaptive response.

Oregon State University

NOTE

[1] This is an adaptation of the author's paper, 'Ballistic Missile Defense and the Possibility of Catastrophic Mistakes,' *IEEE Technology and Society Magazine,* vol. 6, no. 1, March 1987, pp. 4–9.

REFERENCES

[1] U.S. Department of Defense, *Report to the Congress on the Strategic Defense Initiative,* 1985.
[2] U.S. Department of Defense, *Report to the Congress on the Strategic Defense Initiative,* June 1986.

[3] Adam, J.A., and P. Wallich, 'Mind-Boggling Complexity,' first in a series on Star Wars, 'SDI: The Grand Experiment,' *IEEE Spectrum*, vol. 22, no. 9, pp. 36–46, Sept. 1985.
[4] Fischetti, M.A., 'Exotic Weaponry,' second in a series on Star Wars, 'SDI: The Grand Experiment,' *IEEE Spectrum*, vol. 22, no. 9, pp. 47–54, Sept. 1985.
[5] Adam, J.A., and J. Horgan, 'Debating the Issues,' third in a series on Star Wars, 'SDI: The Grand Experiment,' *IEEE Spectrum*, vol. 22, no. 9, pp. 55–64, Sept. 1985.
[6] Carter, A.B., 'Directed Energy Missile Defense in Space,' *Office of Technology Assessment,* Congress of the United States, Washington, DC, April 1984.
[7] Fletcher, J.R. (Study Chairman), *The Strategic Defense Initiative,* Defensive Technologies Study, unclassified summary, Department of Defense, April 1984.
[8] Fletcher, J., Study Chairman and B. McMillian, Panel Chairman, 'Battle Management, Communications, and Data Processing (U),' *Report of the Study on Eliminating the Threat Posed by Nuclear Missiles (U)*, Feb. 1984.
[9] U.S. Senate Hearings, *Strategic Defense and Anti-Satellite Weapons,* hearings before the Committee on Foreign Relations, United States Senate, 98th Congress, April 25, 1984.
[10] Bracken, P. *The Command and Control of Nuclear Forces,* New Haven, CT: Yale University Press, 1983.
[11] Steinbruner, J., 'Launch under Attack,' *Scientific American,* vol. 250, no. 1, pp. 37–47, Jan. 1984.
[12] Wallace, M.D., B.L. Crissey, and L.I. Sennott, 'Accidental Nuclear War: a Risk Assessment,' *Journal of Peace Research,* vol. 23, no. 1, pp. 9–27, March 1986.
[13] Zraket, C.A., 'Strategic Defense: a Systems Perspective,' *Daedalus,* vol. 114, no. 2, pp. 109–126, Spring 1985.
[14] Parnas, D.L., 'Software Aspects of Strategic Defense Systems,' *American Scientist,* vol. 73, no. 5, pp. 432–440, Sept.-Oct. 1985.
[15] Cohen, D. (Study Chairman), *A Report to the Director, Strategic Defense Initiative,* Eastport Study Group, Summer Study 1985, Dec. 1985.
[16] Keyworth, G.A., 'The Case for Strategic Defense: An Option for a World Disarmed,' *Issues in Science and Technology,* vol. 1, no. 1, pp. 30–44, Fall 1984.
[17] Jastrow, R., 'Interview,' *Omni,* vol. 7, no. 12, pp. 61–67, Sept. 1985.
[18] Bella, D.A., 'Nuclear Deterrence: An Alternative Model,' *IEEE Technology and Society Magazine,* vol. 6, no. 2, pp. 18–23, June 1987.
[19] Carnesale, A., Senate Testimony, *Strategic Defense and Anti-Satellite Weapons,* Hearings before the Committee on Foreign Relations, United States Senate, April 25, 1984.
[20] Bowman, R.M., 'Star Wars and Security,' *IEEE Technology and Society Magazine,* vol. 4, no. 4, pp. 21–23, Dec. 1985.
[21] C. Perrow, *Normal Accidents,* Basic Books, New York, 1984.
[22] Waldrop, M.M., 'Resolving the Star Wars Software Dilemma,' *Science,* vol. 232, pp. 710–713, May 9, 1986.
[23] Anonymous, 'Fact Sheet Outlining United States Space Policy,' *Public Papers of the Presidents of the United States – Ronald Reagan,* Book II, 1982, U.S.

Government Printing Office, Washington, DC, 1983.
[24] Smith, J.R., 'A Risk Reduction Center Gains U.S. Support,' *Science,* vol. 231, pp. 107–108, Jan. 10, 1986.
[25] Bella, D.A., 'Fault Tolerant Ballistic Missile Defense: Defining Constraints,' *IEEE Technology and Society Magazine,* vol. 1, no. 3, pp. 22–25, Sept. 1988.
[26] U.S. Defense Science Board Task Force (R.R. Everett, Chairman), *Defense Science Board Evaluation of Status and Plans of Strategic Defense Initiative,* Congressional Record, vol. 133, no. 116, pp. H6292–H6294, July 14, 1987.
[27] Foley, T.M., 'Congressional Report Alleges SDI System Prone to Failure,' *Aviation Week and Space Technology,* p. 19, May 9, 1988.
[28] Strategic Defense Initiative Organization, *Report to Congress on Strategic Defense System Architecture,* Washington, DC, Jan. 1988.
[29] Bella, D.A., 'Organizations and Systematic Distortion of Information,' *Journal of Professional Issues in Engineering,* ASCE, vol. 113, no. 4, pp. 360–370, Oct. 1987.
[30] Bella, D.A. and W.S. Overton, 'Environmental Planning and Ecological Possibilities,' *Journal of the Sanitary Engineering Division,* ASCE, vol. 98, no. SA3, pp. 579–592, June 1972.
[31] Winner, L., *Autonomous Technology,* The MIT Press, Cambridge, MA, 1977.

EDWIN LEVY

JUDGMENT AND POLICY: THE TWO-STEP IN MANDATED SCIENCE AND TECHNOLOGY

INTRODUCTION

The title of my paper may appear cryptic, so I shall begin with a rough explanation. By 'mandated science' I mean the work of scientists and technologists in the context of bodies mandated to make recommendations or decisions of a policy or legal nature. The clearest examples of such bodies are regulatory agencies, expert commissions, standard setting organizations, and the courts. My colleagues, Liora Salter and William Leiss, and I are involved in a research project in which we are trying to describe and analyze 'mandated science' *via* case studies centered on four substances that have been the subject of extensive investigation by a variety of mandated bodies. The substances are lead, fluorides, chlorophenol (the wood preservative), and the pesticide Captan.[1]

The 'two step' mentioned in the title is, crudely, the attempt to make a sharp conceptual and even procedural distinction between, on the one hand, scientific and, on the other hand, political/ethical considerations in public policy making. One of the clearest examples of this attempt can be found in William Lowrance's book, *Of Acceptable Risk*. There he argues that the *measurement of risk* is an empirical, scientific, probabilistic, and objective activity (Lowrance, 1976, pp. 8 and 75–76). In contrast, *judgments of safety*, i.e., judgments about the *acceptability* or *management* of risk are 'a matter of personal and social value judgment' (p. 8); they are subjective and political/ethical (p. 76). Let me emphasize here that attempts to employ this distinction are definitely not confined to the academic arena. Some governments, e.g., West Germany, have explicitly built this distinction into their policy-making procedures; many others have done so implicitly; and still others are contemplating doing so. Agencies in the United States and Canada are in the latter two categories.

I shall now give a more detailed account of 'mandated science'; then I shall make a few remarks about standard setting and one of its primary components, risk assessment. The bulk of the paper will be devoted to showing that in spite of the initial plausibility of the 'two step,' such an approach is often misguided and obfuscating. My position is that while there may be practical reasons to subdivide policy deliberations involving science and technology, it is almost always a mistake to think that scientific factors can thereby be isolated from value considerations.[2]

Unfortunately, such a thesis can all too easily be construed as yet another instance of flogging a straw man who believes in value-free inquiry. I am trying to do something more; I am attempting to show just where and why there are significant philosophical questions in the mandated area. If, for example, a philosopher reading Keepin and Wynne's (1984) critique of a major global energy model – one constructed at the International Institute of Applied Systems Analysis – might conclude that the issues are exclusively technical and sociological, the model builders used questionable techniques and some were allegedly biased towards particular outcomes. However, I believe that a full analysis of such technological tools also requires studying the underlying epistemological issues – e.g., what are the grounds for claiming that a simulation model deserves to be believed? – and deep investigation of the political and ethical terrain in which such technological devices are constructed and employed.

I. MANDATED SCIENCE

In recent decades several commentators have suggested that there is something distinctive going on at the interface of science and public policy. Alvin Weinberg (1972) coined the term 'trans-science' to denote issues that appear to be amenable to scientific resolution but are not; Wessel (1980) has spoken of 'socio-scientific disputes'; Grobstein (in Crandall and Lave, 1981) remarks that there are three branches of science: pure, applied, and policy science; and Jerry Ravetz (1984), Helga Nowotny (1984), and others have recently made similar suggestions.

Since I do not have time to argue in detail that there is a whole new category of science and technology, I shall merely mention a few key features of and questions about mandated science.

(a) First, 'mandated science' is not to be understood merely as 'mission oriented' science. Rather, a significant proportion of the scientific data and personnel that are brought to bear in a mandated context – for concreteness, think of an expert committee struck by a regulatory agency to advise on standards related to the use of a pesticide – come from ordinary, disciplinary scientific research. That is, the studies are carried out as part of an ongoing research programme within a scientific discipline – e.g., immunology or biochemistry – and as such they are peer reviewed, published, read, and judged in accordance with the current scientific and academic outlook of the discipline concerned. However, I believe that such data and papers must often undergo a kind of translation process when they enter the mandated arena. This is because the emphases and very bases of judgment are likely to differ significantly in the two contexts. To get a hint of the differences, think of the contrast between a scientist delivering a paper at the annual meeting of her disciplinary sub-specialty and the same scientist appearing as an expert witness in a lawsuit. There is far more than a difference of style involved here; standards of proof, nature of evidence, concepts of cause and effect, control of the situation are some of the basic categories of contrast.

(b) Second (and somewhat at odds with the previous remarks), a great deal of the material regarded as scientific in the mandated arena does not originate as orthodox disciplinary research. Instead, much of it has been undertaken by industry, government, and even academic investigators because of regulatory and other legal requirements, e.g., to test the efficacy and safety of drugs, pesticides or food additives. In the course of doing interviews for our case studies, we have often asked knowledgeable people about the percentage of non-disciplinary research at the basis of regulations in their fields; the lowest response was fifty percent. Clearly this is a matter for further research. Such studies certainly employ scientific methodology in their use of controls, screening mechanisms, and statistical analyses, but they differ from orthodox disciplinary research in that their intended and actual audiences importantly include non-scientific actors in the mandated arena. In saying this, I am not denigrating these investigations; I am merely pointing to a possible contrast between such studies and what many of us academics consider to be paradigmatic scientific activities.

(c) Third, there is the question whether what takes place in the mandated arena is itself a scientific activity. I suspect there is a wide spectrum. At one end there is the narrowly based scientific advisory committee undertaking what is essentially a literature review in a field common to all of its members. The output of such a committee differs but little from a jointly written review article. At the other end of the spectrum is the committee made up of experts from a wide variety of fields attempting to report on issues from an even wider area than is represented by the composition of the committee. In this latter case, it is difficult to consider the committee's deliberation to be a straightforward extension of what individual or groups of researchers do when making decisions about how to proceed in their research. That is, although there are certainly some committees whose deliberations are barely distinguishable from what individuals or small groups encounter in the normal course of scientific work, there are also committees that are engaged in very different sorts of deliberations. At this point I am not at all certain about the criteria to employ in distinguishing scientific deliberations from non-scientific ones. What is clear to me is that many of these deliberations are unlike those which confront scientists and technologists undertaking research work.

Surely these three features neither provide an exhaustive characterization nor demonstrate that there is indeed a distinctive activity called 'mandated science.' However, we should remember how long people have been trying to produce adequate criteria to distinguish pure from applied science and how inconclusive their efforts have been. As for mandated science, what has to be done is to study in much greater depth how science and technology are handled in the policy arena. Whether or not we conclude that there is a distinctive activity taking place is much less important than analyzing and perhaps contributing to the improvement of what does take place.

II. SCIENCE AND STANDARDS: A MISLEADING MODEL

A candidate for being an appropriate model for the relationship between scientific information and the setting of standards could be the story of the calendar. (See R. F. Legget, 1971; and T. S. Kuhn 1957.) The calendar currently in general international use is a variant of the one developed under the auspices of Julius Caesar in 46 B.C. Before then the system in use had the length of the year as 355 days. It soon became clear that the civil year was about 10 days short, so an extra month was inserted in February from time to time at the discretion of local political authorities. In order to straighten out the situation, Julius Caesar consulted the astronomer Sosigenes and the latter developed a calendar based on a year with 365 and 1/4 days.

Except for a minor adjustment in 8 B.C., said to be necessitated by an error in carrying out Julius Caesar's instructions, that calendar remained in force for fifteen centuries. In 1581 Pope Gregory XIII decreed that the leap year day added to February is to be dropped in every century year except those divisible by 400 (so 1900 was not a leap year, but 2000 will be). What occasioned that decree was the observation, among others, that religious holidays were getting out of synchronization with the seasons. That discrepancy could be explained by the astronomers who had discovered that the length of the year is about 11 minutes 14 seconds shorter than 365 1/4 days. This difference amounts to 1 day in 128 years, so over 15 centuries the effect, an error of about 10 days, was noticeable. Incidentally, not all jurisdictions went along with Pope Gregory; Great Britain did not effect a change until 1752. By then the error was 11 days and when the calendar reform was brought in there was some opposition based on the view that people's lives were being shortened – the slogan was 'Give back our fortnight.'[3]

The salient features of the calendar story are that scientists are called upon to provide the facts – tell us, please, what is the actual length of the year? – and political or other authorities decide how to incorporate those facts into standards. The views of scientists may be of some assistance in the latter decision, but here administrative convenience, political expediency, and social considerations dominate the scene. Now there are undoubtedly some standard-setting situations for which this scenario is apt, but today the standards that concern us most do not fit this mold.[4]

III. RISK ASSESSMENT AND SETTING STANDARDS: MEASURING RISK

A much more realistic picture of the tasks addressed in mandated science today can be obtained by examining the area of risk assessment. I shall divide my examination of this activity into the parts that some recommend as appropriate, namely, the measurement of risk and the acceptability (or management) of risk. My view is that this partition must be judged unacceptable if, as proponents intend, the measurement of risk is regarded as an objective, scientific undertaking and value factors are relegated to the latter.

But first a caveat. There are some recent characterizations of the area that are more elaborate than Lowrance's. For example, in the U.S. National Academy of Sciences volume, *Risk Assessment in the Federal Government: Managing the Process* (1983), the authors distinguish between risk management and risk acceptance. They then acknowledge that policy decisions bear on both of these categories. Although this work certainly constitutes an advance, the basic issues I am addressing are not resolved. So I shall continue to target my comments at the simpler distinction between risk measurement and risk acceptability; I believe my criticisms apply to the more elaborate schemes.

To begin with, I shall examine the measurement of risk under the assumption that it can be clearly distinguished from judgments about the acceptability of risk. In the next section I shall challenge that assumption.

Consider Figure 1 (which I have placed in an appendix), 'Components of Risk Assessment for Carcinogens: A Framework.' The table is drawn from the working papers prepared for the U.S. National Academy of Scicnces (1983) committee responsible for writing the volume mentioned above. The committee's staff conducted a survey of 'scientists and social scientists who were knowledgeable about carcinogenic risk assessment and its uses in policy' to ascertain where each of the components falls along the scientific-value spectrum.[5] The results of the survey are that none of the components are rated as being purely scientific and six are seen as being pure value judgments. (See components 5, 24, 30, 32, 34, and 36; a 'V' appears beside these in the figure.) Neither the list nor the ratings are definitive.[6] However, I believe that both are reasonable attempts to delineate and rate the factors that must be taken into account when estimating the cancer risk associated with a given substance.

I do not have space to go over the table in detail, so I shall simply lay out my main claim. If the two-step approach were correct, then you would expect that number 36 in the table, the judgment of the acceptability of risk, would be in a class by itself; the elements preceding it would be labeled 'scientific,' and number 36 would be considered 'value.' Instead, what the table suggests, and what I believe is indeed the case, is this:

[1]   *Many decisions required for the measurement of risk are value-laden in the same way as are judgments about the acceptability of risk.*

I offer two argument sketches directly relevant to this claim. (In section V I shall mention some more general conceptual considerations that support the claim.)

(i) Feedback argument: One could attempt to show that estimates of the acceptability or management of risk must be taken into account when measuring the risk.

I believe that there is substantial empirical evidence supporting [1]. For example, Gillespie *et al.* (1982) did a comparative study of decisions in the U.S. and U.K. concerning the registration of the pesticides Aldrin/Dieldrin. Although the scientific evidence bearing on the decisions was the same, the U.S. chose to suspend registration and the U.K. decided to continue using the substances. There are of course many considerations contributing to the different decisions, but one of the prominent factors was the different orientations of the main scientific actors in the two cases. In the U.S., one of the chief scientific witnesses viewed the issue from the perspective of health. He believed that a substantial proportion of cancers have an environmental origin; that this situation is unacceptable; and thus that where decisions could be made to avoid environmental risk, they should be so made. In contrast, one of the main scientific participants in the U.K. stressed world food shortages. He argued that pesticides contribute significantly to food production and that their risks are greatly overestimated.

The point of this and many similar examples is not that scientists are biased in some tawdry sense – though undoubtedly there are examples of this as well. Rather the point is that estimating the magnitude of risks is a process fraught with complexity and uncertainty, and that in such circumstances one's general position about the gravity of the situation cannot help but become an operative premise in one's estimates of the risks.

In this case some of the main issues concern criteria for carcinogenicity, and in particular the problem of estimating low dose effects. Such effects are of course notoriously difficult to study either by animal tests or by epidemiological research. With respect to cancer, there are divergent, and strongly held, views about the mechanisms of carcinogenesis. According to the 'threshold' view, there are doses below which carcinogenic substances have no effect; thus, the extrapolation of dose-response curves should show zero effects at low doses. In contrast, according to the 'trigger' view, cancer can result from a single hit from such substances; thus extrapolation of dose-response curves should reflect the possibility of effects at low doses.

In a report on dioxins, an expert committee compared threshold approaches with trigger calculations, where the latter regarded a 'virtually safe dose' (VSD) as one which produced one cancer in a million individuals. The result was that threshold estimates sanctioned

... an exposure level some 6 to 30-fold greater than the mean VSD at [a one-in-a-million] risk level. ... A 6 to 30-fold difference between the two methods of extrapolation of the carcinogenicity data is well within the uncertainties in the data and the methodologies involved. From a practical point of view the two procedures give essentially the same result with the exception that the threshold hypothesis assumes no risk of cancer below the acceptable intake level (Health and Welfare Canada, 1983, p. 36).

For some agencies in the United States and elsewhere, this matter has been 'settled' by a policy decision. However, for other jurisdictions and for many other kinds of risks, which model to adopt is a matter of judgment within mandated bodies. Clearly, one's judgment about the acceptability of the risk can be a factor in deciding which model to adopt.[7]

(ii) The second argument involves showing that some or many of the 'V' assignments in schemes like figure 1 are indeed justified. By way of example consider item number 32 in the table, 'estimating intake,' an aspect of estimating exposure. Consider this summary of an expert witness's testimony in the recent Nova Scotia herbicide case; the author of the summary is the presiding judge, Justice J. Nunn:

> The witness and several others did a test on themselves to study skin absorption by strapping a one square foot cloth, saturated with a normally used concentration of [2,4,5-T] spray, to their thighs and leaving it for two hours. The result was that the skin is a very effective barrier, for just over two milligrams of the chemical migrated into the body and this amount is one-thousandth of a maximum reported safe level of 2,4,5-T on a daily basis for life. This led him [the witness] to the conclusion that even the applicators of the spray are under no risk unless they are spraying over their heads with their mouths open (Palmer *et al.* v. Nova Scotia Forest Industries, 1983, p. 199).

Let me hasten to say that such evidence would not stand up in many other arenas of mandated science, not even in other courts.[8] But even if this is an extreme case, many of its elements are common enough: estimates of real-world exposure extrapolated from very different conditions; the use of fancy terminology and quantification to disguise what are really crude estimates; and, even more important, predictions of human behavior.

The last mentioned factor deserves emphasis. In assessing exposures, a great deal of attention has been paid to features amenable to analysis by the physical and biological sciences: atmospheric conditions, soil permeability, degradability, breakdown products, and the like. Increasingly, it has been recognized that one of the weakest aspects of these calculations is the estimation of what likely will or will not be done by people: pesticide applicators, consumers, reactor technicians, etc. In these circumstances, sometimes the real experts are the farm workers and hands-on users. This is the conclusion to be drawn, for example, from experience in the U.K. with regulation of 2,4,5-T. The recommendations of the scientific experts on the Pesticides Advisory Committee were based on prescribed usage. These estimates and the committee's recommendations subsequently had to be substantially revised in light of testimony concerning actual production and usage of the pesticide (Wynne, 1983, p. 10). In the present context, the moral to be drawn is that judgments about exposure levels, among other things, require extensive assessments of social behavior. Such assessment can come from practitioners and/or from social scientists. Frequently it comes from neither; the assessments are based on seat-of-the-pants social science carried out by physical and biological scientists. However, no matter who does it, there seems to be considerable opportunity for values to play a significant role. Thus the 'V' rating of this component of measuring risk seems amply justified.

IV. RISK ASSESSMENT AND SETTING STANDARDS: ACCEPTABILITY OF RISK

There is indeed something commonsensical about viewing questions concerning the acceptability or management of risk as being subjective matters of personal or social value judgment. After all, if an individual is informed about the probabilities and natures of risks he is likely to encounter in following various courses of action, then the individual's choice depends heavily on what level of risk he is willing to tolerate. Of course other factors must be considered such as the balance between likely costs (including risks) and benefits. Still, at the end of the day personal preference must play a non-negligible role. The trouble with this intuition is that especially when we consider social decisions, as I am in this paper, a great deal must take place between the receipt of risk measurements and the end of the day. For example, risk-benefit and/or cost-benefit analyses are thought to be both scientific (or quasi-scientific) activities and part of the process of deciding on the acceptability of risks. Thus my second major claim:

[2]   *There is science involved in judging acceptability and it too can be heavily value-laden.*

Virtually everyone agrees that scientific considerations play a role in judging the acceptability of risk. Lowrance, for example, contrasts the scientific, value-free activity of measuring risk with the value-laden process of judging acceptability; but, far from denying that science is involved in the latter, he is at pains to show that it is. So the contentious part of [2] is the second point, the view that the science that is involved is value-laden.

A natural candidate to use in support of this claim is risk-benefit or cost-benefit analysis. Few topics in the policy arena have received so much attention as these. (See, e.g., Swartzman and Croke 1982; Levy and Copp 1983; Copp and Levy 1981; and others.) I think it is clear that these would establish the claim, but I think it instructive to look elsewhere.

Consider, for example, the notion that a risk is acceptable if it is as low as technically feasible. (This is one of the tests of acceptability mentioned by Lowrance.) How is technical feasibility established? A useful illustration is the issue in Canada of the threat to fish posed by chlorophenol, a wood preservative, that enters the ocean from storm drains of sawmills.[9] That is, in British Columbia and the Pacific Northwest, wood treated with chlorophenol (or chlorophenate) is stored outside and the considerable rainfall washes some of the chemical into the ocean. In the U.S., this problem is handled by establishing water quality criteria and effluent standards. In Canada, the approach is to ascertain the degree to which technically and economically feasible devices can clean the runoff and use that as a guide to setting effluent standards. Systems employing carbon filtration and reverse osmosis are under investigation by government laboratories, and presumably the private sector will or could do some of the R & D. The point is that the acceptability of the risk is tied to what could be regarded as a scientific or technological determination of feasibility. However, notice that the very concept of 'feasibility' can be heavily value-laden; indeed, it is a conceptual cousin to 'acceptability.' Will the search for a feasible option include building sheds for treated wood? Using an alternative chemical? How, when, and where does economic feasibility interact with technical feasibility? The point of these questions is to show that what can look like a strictly technical question, and what may approach being a purely technical issue for an engineer working on flow design for a filter system, turns out not to be merely technical in the policy arena.[10]

This example may seem too narrow to support a general claim, but in the U.S. and in Canada most regulatory statutes make reference to feasibility or cognate terms such as 'as low as reasonably achievable' or 'best available practice.' Thus I think similar examples can be found in the areas of occupational health and safety, radiation protection, hazardous waste disposal, consumer protection, and elsewhere.

V. UNDERLYING CONCEPTUAL ISSUES

There are some fundamental conceptual considerations that if explored in depth would explain why there is no neat division between scientific and socio-political issues.

(i) The effects that are under investigation take place in messy, real-world conditions. This point is so obvious that its deep significance can be missed. In many respects, the birth and development of modern physical science integrally involves the assumption that the entities under investigation can be treated as isolated systems; what physical and other scientists are discovering in the mandated arena is that that assumption does not hold.

(ii) A point closely related to (i) can be called the Socratic Paradox: the more we know, the more ignorant we become. In the current context, the paradox is related to the rise of ecological considerations and the demise of the isolated system. We now know, for example, that chemicals injected into eco-systems tend to have cascading effects, at least some of which are likely to be undesirable. However, we seldom know either the magnitudes or the interactive, synergistic features of many of these effects. Yet the political and legal systems daily call upon scientists for precise answers. The more we learn about effects, the more we know needs to be learned.

(iii) Mandated science is an activity that takes place under a cloud – or perhaps within a fog bank – of uncertainty. It is tempting to say, for example, that the reason values enter into the measurement of risk is that the scientific data are uncertain. That is indeed the case, but if we stop with that statement we conceal more than we reveal. For there are many kinds of uncertainty. In the case of the year length, for example, although at times astronomers may have regarded the data as uncertain, they could establish a fairly uncontentious narrow range in which the correct figure must fall. In contrast, in measuring risks, scientists' conclusions are often qualitatively different. Consider this comment on the saccharin controversy:

... One model predicted 5 cancer cases per million persons exposed to saccharin, another predicted 1200, and [another], offered by an industry group, predicted only one death per one billion persons.[11]

(iv) The viability of a distinction between 'data' and 'interpreting data.'

The question whether there is a legitimate distinction between facts and values and the question whether there can be a theory-free observation base are two hoary philosophical problems. Here I cannot address these issues in detail, but in this and the following subsection I would like to show how versions of these problems arise at the foundations of mandated science, and actually become acute when attempting to integrate data from a variety of disciplines.

One thing contributing to the intuitive plausibility of the two-step approach is the idea that there is a rather clear distinction between, on one hand, 'facts' or 'data' and, on the other hand, interpretation of facts and data. We often hear for example that the data are not in contention, but the interpretation of the data is a matter of dispute. I suspect that there are often disputes about data; furthermore, the disagreements I am alluding to are not concerned with fraud or deception. These disagreements occur especially in such fields as epidemiology and applied toxicology. What distinguishes these fields is that what they regard as data are, from the point of view of more 'basic' disciplines, already interpreted to a significant degree. However, this is to jump ahead of the story. I first wish to show that even within a basic discipline, the question, What are data?, can be non-trivial.

In a provocative paper, Trevor Pinch (1984) discusses how scientists report different results from the same experiment. Pinch examines an experiment designed to study solar neutrinos, these being particularly elusive elementary particles. In order to capture some, physicists placed a huge vat filled with perchloroethylene (dry-cleaning fluid) at the bottom of a disused mine shaft. In theory, capture of neutrinos is indicated by the presence of a radioactive isotope of argon. The following are qualitative versions of observations reported (whereas actual reports are quantified):

(1) Splodges on a graph are observed.
(2) Radioactive argon is observed.
(3) Some neutrinos are detected.
(4) Some solar neutrinos are detected.

Roughly speaking, the differences among these claims is that, given the first, plus a number of important auxiliary assumptions and theories, one can 'derive' the second statement.[12] Similarly, the third and fourth can be derived from the preceding ones plus auxiliary hypotheses. Thus the differences among the statements can be characterized by saying that the statements are increasingly 'hypothesis-laden.'[13] Pinch's point is that what is offered as an 'observation' statement will depend on a number of factors, but especially upon the context and the scientists' audience.

It seems to me that exactly the same point that Pinch is making about observation reports can be made about 'facts' and 'data.' Notice that Pinch is not saying nor am I, that there is necessarily anything illegitimate or untoward taking place here. If there is a debate about heavily hypothesis-laden reports, is this a controversy about the data/facts or about the interpretation of the data/facts? I can easily imagine situations in which this question embodies a gratuitous distinction, one that makes little difference. Let us turn to such a case.

As I suggested above, there appears to be a hierarchy of disciplines such that the less basic use as data inputs what the more basic regard as highly interpreted claims. Consider an epidemiological study in which deaths within human populations are regarded as 'data' and the cause of death is the key factor. For example, if one is comparing cancer deaths in the general population to those occurring in a population exposed to a particular chemical, medical records have to be consulted. It has long been recognized that there may be serious problems here – in fact there are now experts, called nosologists, whose specialty includes interpreting and codifying death certificates. A person who has cancer, including even those who have been 'successfully' treated, can develop all sorts of complications. A person whose immune system has been severely suppressed by chemotherapy or by radiation is vulnerable to opportunistic infections. Although at one level it is correct to say that the cause of death for such a person is the infection, at another level (and for other scientists or medical practitioners), the cause of death is cancer. Thus if the study is based on recorded causes of death, these data can be and are legitimately questioned. Would this be a debate about the data or about the interpretation of data? My point is that it could be regarded as either.

The example I have just discussed is admittedly crude. As I have said, epidemiologists have recognized this problem and have taken some steps to avoid it. The problem could be lessened if the basis of the study were incidences of cancer rather than recorded causes of death. But this raises another problem in turn. Both records of cancer incidence and diagnostic reliability are also contentious. What a general practitioner will call cancer and what an oncologist will may differ; so too for an oncologist and a pathologist; and so too for a pathologist and a biochemist. It seems to me that the general point remains: far from there being a pool of facts/data from which scientists draw conclusions in the mandated arena, there is usually a heterogeneous set of claims, at least some of which qualify as data for some disciplines but which would not be so regarded by other more basic disciplines. My intention here is not to impugn these claims; rather, my point is to show that there is at least the possibility of softness at the foundations of mandated science.

(v) *Interpreting data*: The basic issue here concerns the nature of the considerations that are brought to bear in drawing conclusions about data. (We are now assuming that what counts as data is uncontentious.) However, the question is confounded by the fact that 'interpreting data' covers many and disparate activities. I shall mention only two. First, there is the question whether, given particular research papers, the claims made by the authors are adequately supported by the data. Second, there is the question of assessing the significance of the previous conclusions for the wider issues under consideration in mandated science. It seems to me that to get from the first to the second one may legitimately employ ethical principles. Consider the case of dioxins. At one level, there have been studies attempting to discover whether human cells have receptors for some isomers of dioxin, and if so, what is the 'normal' role of those receptors. A second level of interpretation concerns the significance of the previous conclusions for measuring the risks posed by dioxins. Suppose for example that one concludes that the data indicate that dioxin receptors are in human cells, are part of reproductive mechanisms, and could recognize many isomers of dioxin – not just 2,3,7,8-

TCDD. This position coupled with even a moderate principle of caution would result in one's leaning towards accepting higher risk estimates from all spheres and advocating more stringent regulatory action.

It is important to emphasize that this is not merely a psychological point about how some factors affect one's overall perceptions; it is an epistemological point as well. For when one is making decisions under uncertainty – as is invariably the case in mandated science – there are always ranges of estimates and varieties of supportable ways to aggregate and extrapolate them. I see no grounds for an across the board exclusion of ethical principles as legitimate aids for choice.

VI. POLICY CONSEQUENCES

In conclusion, I wish to mention some of the consequences that follow if my central claims are judged correct, or even plausible.

(1) Flag value features: As a bare-bones minimum, scientists in the mandated arena should seek to identify those factors where their judgments are not straightforward, fairly direct interpretations of data. On one hand this point is obvious. On the other hand it is quite tricky, since my main argument has been that it is difficult if not impossible to disentangle factual and value considerations. What I am suggesting here is that mandated scientists should continually pose this question: *Could this issue be decided by an informed citizen who does not possess my technical expertise?* Whenever this question is answered in the affirmative, it seems to me that mandated scientists should at least flag the issue, and at most refuse attempts to resolve it. Clearly the latter course may require revisions to the mandates scientific experts accept. Below I suggest that is a good thing.

(2) Swamping of scientific expertise: I believe that posing and answering the question above would quickly lead to the conclusion that experts in the mandated arena are less essential than most believe. The sorts of issues addressed are so complex that a given member actually acts in her area of expertise a small proportion of the time. Clearly, this difficulty would not be solved by creating larger committees because the underlying problem is to knit together matters from a variety of disciplines and because the greater the number of specialized experts, the more frequently one is not operating in her area of expertise. Some may suggest interdisciplinary experts. But I doubt that in the foreseeable future there can be such persons, for science currently encourages narrowly focussed specialties.

(3) Finally, there is the question of authority. There is an historic tension

between science and technology making contributions to public authority in the form of advice and invention, and their acting as a restraint on public authority by exercising their critical faculties and, at times, their anti-authoritarian ethos. In recent years the advisory role seems to have been in the ascendance. Now it may be time for scientists and technologists to exercise restraint by circumscribing their own role. Simply put, I believe that they should stop making decisions and recommendations that far exceed their competence. This is not to counsel scientists and technologists to withhold service to government; that would be to invite disaster for society and for the institutions of science and technology. I am suggesting that scientists and technologists be more discerning about the tasks they undertake.

University of British Columbia

APPENDIX

Figure 1.

COMPONENTS OF RISK ASSESSMENT
FOR CARCINOGENS: A FRAMEWORK

I. Hazard Identification
   A. Human Data
      1. Weighting Routes of Exposure
      2. Weighting tumor sites
      3. Benign tumors
      4. Strength of Study
      5. Study Results Positive?   {V}
      6. Aggregation
         (pos & neg studies)

   B. Animal Data
      7. Weighting Routes of Exposure
      8. Weighting tumor sites
         ('Suspect' organs)?
      9. Benign tumors
      10. Tumor multiplicity
      11. Weighting tumors in controls

12. Different experimental groups (Most sensitive dose/sex/strain?)
13. Confidence Levels
14. Strength of studies
15. Adequacy of pathology
16. Physiological extenuations (Special metabolic paths)
17. Aggregation (pos & neg studies)

C. Other data
18. Short-term screening assays
19. Confidence levels in short-term tests
20. Structure/activity analysis

D. General
21. Aggregation – weighting human, bioassay & other tests

II. Dose Response (D-R) Assessment
A. Human Data
22. D-R from Epidemiology
23. Extrapolation – Math Model
24. Expression of D-R: 'best estimates' or upper confidence limits {V}
25. Factoring in physiological extenuations

B. Animal Bioassay
26. Aggregating D-R from various tests
27. Math Model for extrapolation
28. Factoring in interspecific conversions
29. Time-to-tumor effects
30. Expression of D-R {V}

    31. Physiological Extenuations

III. Exposure Assessment
    32. Estimating intake: 'gluttony
        scale'; 'worst case scenario' {V}
    33. Relation of target organ to
        total exposure

IV. Expression of Results
    34. Allowance for most
        susceptible individuals {V}
    35. Units of Risk – death,
        years lost, incidence
    36. Acceptable risk {V}

NOTES

[1] Our project, including research for this paper, is supported by the Social Sciences and Humanities Research Council of Canada through its Programme in the Human Context for Science and Technology (Grant No. 499-83-0002). See Salter (1988).

[2] John Perkins (1984) has described his experiences on a committee created by the State of Washington's Pest Emergency Powers Act. The purpose of this Act is to facilitate response to a threat to forestry and agriculture posed by exotic insect pests. The committee has recommended that its task be divided into two phases: the first involves deciding whether there is a threat of sufficient magnitude to warrant emergency action; and the second concerns deliberations about what actions should be taken. My point should not be construed as an objection to this division *simpliciter*. However, I think it indefensible to regard either phase as exclusively scientific and objective. Both phases involve scientific and political/ethical considerations, though in each some determinations may be more amenable to objective analyses.

[3] At first I suspected that this story was apocryphal, but then I recalled that at one time my mother objected to daylight saving time on the grounds that it caused her plants to get too much sun.

[4] Actually, it is arguable that even the calendar story does not fit exactly. Recall that early in the sixteenth century Copernicus was asked to sit on the commission to reform the calendar and he refused because he thought the astronomical data were inadequate.

[5] U.S. National Academy of Sciences, *Working Papers* (1983), p. 91. There is no indication of the number of respondents.

[6] In fact, in U.S. National Academy of Sciences, *Risk Assessment* (1983), the list is expanded to 50 components.

⁷ I am fully aware that the preceding argument sketch is vulnerable to the philosophical objection that descriptive evidence about what is done in practice does not establish the philosophical claim that measuring risks *necessarily* involves judgments about the acceptability of risk. That is, a philosopher could argue that the scientists were simply confused about what premises are legitimate. I believe that I meet that objection in part V, section (v). In addition, I believe that there are some risks that, by their very nature, require estimates of acceptability in order to gauge their extent. For example, fear of risks is itself a risk, so if one is estimating the total risk, one must take into account the perceived acceptability of risks.

⁸ For example, although the Nova Scotia herbicide case (Palmer *et al.* v. NSFI, 1983) has been widely noticed and discussed, there was a tribunal in Canada convened at the same time that considered essentially the same issues and much of the same evidence and reached a different conclusion. This was the Ministers' Expert Advisory Committee on Dioxins convened jointly by Health and Welfare Canada and by Environment Canada (1983).

⁹ This illustration is based on an interview the author conducted with a Canadian government official, November 1984.

¹⁰ I assume that the relevant government agencies have made or will make a judgment whether the most feasible system does in fact protect the fish.

¹¹ See R.A. Liroff, 'Cost-Benefit Analysis in Federal Environmental Programs,' in Swartzman and Croke (1982).

¹² Although we philosophers of science are fond of thinking of derivation as a valid deductive argument, these are merely argument sketches.

¹³ Pinch introduces the concept of 'degree of externality' to characterize the differences. I am not convinced that notion makes things clearer.

REFERENCES

Copp, D. and E. Levy. 'Value Neutrality in the Techniques of Policy Analysis: Risk and Uncertainty,' *Journal of Business Administration,* 13(1981):161.

Crandall, R.W. and L.B. Lave, eds. *The Scientific Basis of Health and Safety Regulation* (Washington D.C.: The Brookings Institution, 1981).

Gillespie, B., D. Eva, and R. Johnston. 'Carcinogenic Risk Assessment in the USA and UK: The Case of Aldrin/Dieldrin,' in B. Barnes and D. Edge, eds., *Science in Context* (Milton Keynes, U.K.: Open University Press, 1982).

Health and Welfare Canada and Environment Canada. 'Report of the Ministers' Expert Advisory Committee on Dioxin' (Ottawa: Government of Canada, 1983).

Keepin, W. and B. Wynne. 'Technical Analysis of IIASA Energy Scenarios,' *Nature,* 312 (20 December 1984):691.

Kuhn, T.S. *The Copernican Revolution* (New York: Vintage, 1957).

Legget, R.F. *Standards in Canada* (Ottawa: Information Canada, 1971).

Levy, E. and D. Copp. 'Risk and Responsibility: Ethical Issues in Decision-Making,' in W. Cragg, ed., *Contemporary Moral Issues* (Toronto: McGraw-Hill Ryerson, 1983).

Lowrance, W. *Of Acceptable Risk* (Los Altos, CA: William Kaufmann, 1976).

McCray, L. 'An Anatomy of Risk Assessment,' in U.S. National Academy of Sciences, *Working Papers* (see below).
Nowotny, H. 'A New Branch of Science, Inc.,' paper presented to International Institute for Applied Systems Analysis International Forum on Science and Public Policy, Laxenburg, Austria (unpublished, 1984).
Palmer et al. v. Nova Scotia Forest Industries, *Canadian Environmental Law Review* 12(6) (1983):157.
Perkins, J.H. 'Medflies, Apple Maggots, Gypsy Moths, and Other Curious Invaders of the Pacific Northwest,' a talk given at the Pacific Northwest Association for Environmental Studies, Victoria, B.C. (unpublished, 1984).
Pinch, T. 'Towards an Analysis of Scientific Observation: The Externality and Evidential Significance of Observation Reports in Physics,' a paper presented in 1984 (forthcoming in *Social Studies of Science*).
Ravetz, J.R. 'Uncertainty, Ignorance, and Policy,' paper presented to IIASA International Forum on Science and Public Policy, Laxenburg, Austria (unpublished, 1984).
Salter, L., Levy, E., and Leiss, W. *Mandated Science: Science and Scientists in the Making of Standards* (Dordrecht: Kluwer, 1988).
Swartzman, D. and K.G. Croke, eds. *Cost-Benefit Analysis and Environmental Regulation* (Washington, D.C.: Conservation Foundation, 1982).
U.S. National Academy of Sciences. *Risk Assessment in the Federal Government: Managing the Process* (Washington, D.C.: National Academy of Sciences, 1983).
U.S. National Academy of Sciences. *Working Papers: Risk Assessment in the Federal Government* (Washington, D.C.: National Academy of Sciences, 1983).
Weinberg, A. 'Science and Trans-science,' *Minerva* 10 (1972):202.
Wessel, M.R. *Science and Conscience* (New York: Columbia University Press, 1980).
Wynne, B. 'Public Perceptions of Risk: Interpreting the "Objective Versus Perceived Risk" Dichotomy,' a paper prepared for the International Institute for Applied Systems Analysis (unpublished, 1983).

PART II

HISTORICAL DIMENSIONS

LEE W. BAILEY

SKULL'S DARKROOM:
THE *CAMERA OBSCURA* AND SUBJECTIVITY

When Richard Rorty declared in 1979 that, 'It is pictures rather than propositions, metaphors rather than statements, which determine most of our philosophical convictions' (p. 12), he was echoing Aristotle's insight, that 'it is impossible even to think without a mental picture.' Despite the antiquity of this insight, long-buried root metaphors still must be repeatedly dug up and scrubbed off. Like the uncovering of long-hidden foundations of classic monuments, the discovery of the root metaphors of established notions calls for a reevaluation of the ideas, from the bottom up. To find a concept's actual significance, Stephen Pepper (1942) said, 'Trace it back to its root metaphor' (p. 114). The modern supposition of subjectivity is one such monumental concept, and the common *camera obscura* is one of its buried root metaphors.

The *camera obscura* is simply a dark room with a small hole permitting light to enter. If a white screen is set up opposite the hole, the external scene will be projected on the screen upside down. Place a lens in the hole and the picture can be focused and inverted. The Gernsheims's (1969) research has shown that as early as the Middle Ages the *camera obscura* was used to view eclipses and to study light. Later, Roger Bacon (ca. 1250) used a mirror in a *camera obscura*. Leonardo da Vinci (ca. 1508) built a small *camera obscura* to study perspective drawing. René Descartes, in 1637, put an ox eye in the hole to study focus. John Locke, in 1690, wished the moving pictures in the *camera obscura's tabula rasa* would stand still. Sitting in their dark chambers, such thinkers were fascinated with the way it inevitably raised basic questions about the structure of the human eye, the reliability of perception, the nature of human memory and understanding. Small portable darkrooms were common by the eighteenth century, when tricksters could be found hiding in carriages secretly viewing others on an interior screen. Dramatic theatrical applications of the *camera obscura* originated as early as the Renaissance, when in 1589, dinner guests of Giambattista della Porta were alarmed to see a mysterious moving picture on a dark room's wall, and when in 1613, carnival patrons were terrified to see an apparition of a devil parading across a dark room's wall. The *camera obscura* was a widely known device for centuries before Daguerre inserted photosensitive plates into one in 1839, and invented the photographic camera. But what historians have not noticed is that throughout its history the *camera obscura* has quietly but significantly functioned as a guiding root metaphor for our modern view of the soul.

Whenever we sit down in a dark room to watch a movie, a slide show, or television, we enter the imaginative world of the *camera obscura,* the dark room of the soul. Apart from the contents of its projected dramas, the very structure of the dark room itself has influenced our modern concept of the subjective psyche in unacknowledged and problematic ways. The *camera obscura,* dating back to the Renaissance, has long been a root metaphor, a largely unconscious guiding image that lends plausibility to the narrow, alienating, post-Cartesian idea that the psyche is a purely internal entity contained in a little black box, the dark room of the skull. Gilbert Ryle (1949) rightly emphasizes the metaphoric nature of 'Descartes' Myth,' the mind/body dualism. But resistance to overcoming the dualism may be due to another metaphor lurking in its depths, the *camera obscura*.

The influence of this simple machine on thought has been all the more effective because it has been largely unconscious. Though on occasion explicit, the fantasy that the psyche is in a container is most often a dreamlike latent image hiding under the surface of the literature of the *camera obscura*. The hidden analogy of the psyche to the *camera obscura* grew slowly. It was first nourished in the Renaissance image of individualism and perspective art, then it shaped Locke's epistemology of representation and the *tabula rasa*. Later it fed into the nineteenth-century psychology of ego-subjectivity, strong in Feuerbach (1841) and Freud (see 1974).

The effect of this buried metaphor was subtle but sure. It gave its structure to a new way of imagining the psyche as a contained, internal entity, ontologically divided from the external world, communicating with that alien space only through restricted channels. This communication kept the image from lapsing into solipsism. But by contrast with the ancient vision of the microcosm reflecting the macrocosm, the contained, internal darkness of this root metaphor greatly reduced the picture of the soul's participation in the world. The *camera obscura* began as an experimental model for the eye and became a ruling metaphor for the mind. By offering a way of picturing the Cartesian inside *cogito* with a sensory channel admitting pictures from the outside *extensio,* the image of skull's darkroom shifted from a suggestive experimental analogy to a concealed methodological paradigm.

It would be improper to say that this image contributed to the idea of subjectivism in any causal fashion. Images offer organizing patterns to conscious data, but cannot be said to *cause* ideas. On the contrary, the idea of cause and effect relations itself has its own root metaphors, such as the chain of being and bouncing billiard balls. Such images govern ways of seeing, even establish orthodox procedures of thinking. James Hillman (1975) sees such root metaphors as archetypal images, and he stresses their 'emotional possessive effect, their bedazzlement of consciousness so that it becomes blind to its own stance' (p. xiii). Guiding images unconsciously organize patterns of understanding, but they are not immutable. At its best the therapy of ideas in each age brings to light these structures, thereby freeing thought to seek better root metaphors.

Contemporary awareness of this situation is sketchy. In France, Sarah Kofman has pointed out the impact of the *camera obscura* in her 1973 book *Camera obscura de l'ideologie*. The root metaphor of the darkroom receiving projections inward, she shows, has been used to support two contradictory Western ideologies. Leonardo da Vinci and Jean-Jacques Rousseau argued that the representation of the external world inside the *camera obscura* supports the view that the pictures in the blank screen of the mind accurately represent 'objective reality' (representationalism). But conversely Marx, Nietzsche, and Freud saw the image of psyche modeled on the darkroom as evidence that unreliable delusions are projected into the blank screen of the mind's inner chamber. To them the image shining on the mental screen are delusory inversions of social relations and conscious knowledge, whose origin has been forgotten. Kofman comments:

All this happens then, as if the key to the dark room had been thrown away and the idea abandoned inside, a prisoner, constrained to turn narcissistically around itself (p. 18).

In both theories the presupposed paradigm is the darkroom that receives projections from outside onto its blank screen. But the dispute is over whether to trust or distrust the incoming sensations.

Kofman has begun to unveil the hidden influence of the *camera obscura* on these Western ideologies. Yet a deeper level of the exploration of this basic root metaphor itself must be uncovered. Not only has the darkroom unconsciously guided central Western ideas of objective representation and delusory distortion, but underlying these ideologies lies another, hardly criticized ideology, that of subjectivism. This presupposition has pictured psyche isolated in a subjective container that has become a mental prison. In order to fully deconstruct the representation/delusion debate, there must first be a critique of its foundation, the *camera obscura* paradigm itself.

Owen Barfield has begun this task in *The Rediscovery of Meaning* (1977).

He reminds us that archaic cultures did not begin with the paradigm of the mind as an isolated unit confronting a separate 'objective' world. The 'onlooker' stance is not natural. Rather, mankind has had to 'wrestle his subjectivity out of the world of his experience by polarizing that world gradually into a duality. And this is the duality of subjective-objective, or outer-inner, which now seems so fundamental' (pp. 16–17). The *camera obscura* was instrumental in bringing this mental construction about, Barfield says. The Renaissance *chambre obscure*, and its descendant the nineteenth century photographic camera, have made us into a civilization of 'Camera Man' (p. 76). It has long offered an imaginative structure leading to the vision of the soul shut up in a dark box, peering out at the world from a stance of isolated, lonely, alienated subjective egos. For Barfield, the age of Camera Man is not an inevitable development but a distorting aberration. But Barfield's initial insight needs historical confirmation.

THE ARTIFICIAL EYE OF THE SOUL

The first recorded use of the *camera obscura* was its use as an instrument of optical research by the Arabic scholar Ibn Al-Haythan (965–1039), known as Alhazen (Sabra, 1970). By the thirteenth century, the *camera obscura* was commonly used in England and France for viewing solar eclipses (Gernsheims, p. 18). Roger Bacon (ca. 1220–1292), according to the same source, described the use of the *camera obscura* with an external mirror:

Mirrors may be so arranged that we may see whatever we desire and anything in the house or in the street, and everyone looking at those things will see them as if they were real, but when they go to the spot they will find nothing (p. 166).

Bacon's early note that the inwardly projected images seem very 'real,' but actually are nothing, sets up the longstanding debate noted by Kofman (1973). Are the images seen in the *camera obscura* trustworthy or delusory? Obviously this question does not remain a simple question about a room used for studying light. It welcomes the parallel with the eye. If light behaves in the eye like it does in the darkroom, can such sensations be trusted or not? This ancient debate gained a tool for experiment in the *camera obscura*.

Leonardo da Vinci made the earliest recorded explicit comparison of the *camera obscura* to the eye, calling it an artificial eye:

When the images of illuminated bodies pass through a small round hole into a very dark room, if you receive them on a piece of white paper placed vertically in the room at some distance from the aperture, you will see on the paper all those bodies in their natural shapes and colors, but they will appear upside down and smaller. ... The same happens inside the pupil (Gernsheims, p. 19).

Leonardo sketched the earliest surviving drawing of a *camera obscura*, 'to bring the image of a crucifix into a room.'

Leonardo saw accurate representation of external images on the internal screen, and this allowed him to compare the *camera obscura* to the eye, trusting the incoming sensations. By the mid-sixteenth century the *camera obscura* was known in England, France, Holland, Germany, and Italy (Gernsheims, pp. 18–19).

By Descartes's time the artificial eye was considered such an important test of the question of the reliability of sensations that he discussed an experiment in *La Dioptrique* in 1637:

If a room is quite shut up apart from a single hole, and a glass lens is put in front of the hole, and behind that, some distance away, a white cloth, then the light coming from external objects forms images on the cloth. Now it is said that this room represents the eye; the hole, the pupil; the lens, the crystalline humor – or rather, all the refracting parts of the eye; and the cloth, the lining membrane, composed of optic nerve-endings (1913 ed., vol. 6, pp. 114 ff.).

Fitting an eyeball into the room's opening to demonstrate the eye-dark room analogy, Descartes discovered the wonder: 'You will see (I dare say with surprise and pleasure) a picture representing in natural perspective all the objects outside.' Descartes was forced by the *camera obscura* to admit the accuracy of the representations of external bodies on the blank screen. Locke would later take this to indicate the reliability of visual perception. But Descartes insisted that despite their apparent reliability, the images on the screen do not overcome the problem of delusion, which could be located further back in the brain's passages.

Leonardo's analogy between the *camera obscura* and the eye was the first step in setting up a new mental map of the soul. The question whether the images projected into the darkroom are veridical or delusory was not solved by the *camera obscura* itself. Even to explore this question the darkroom must enlarge to become a metaphor for the soul, because it raises unavoidable interpretive issues. The *camera obscura* is not a neutral tool revealing any supposed 'objective truth' about perception or epistemology. It is a metaphor, a map, a model for a way of imagining, it is 'something like' the Eye of the Soul.

THE BLANK SCREEN OF THE SOUL

John Locke realized this and expanded the analogy of the *camera obscura* to let it represent the whole of human understanding. His response to the colorful images on the white surface opposite the aperture was just the opposite of Descartes's. He wholeheartedly believed the images, as if they were trustworthy representations of the external world. Because of his faith Locke made the great leap beyond the analogy of the *camera obscura* as an artificial eye. He awarded this dark closet the honor of becoming an explicit key metaphor for human understanding. In the *Essay Concerning Human Understanding* (1690), he modestly proclaimed:

Dark Room. – I pretend not to teach, but to inquire, and therefore cannot but confess here again, that external and internal sensation are the only passages that I can find of knowledge to the understanding. These alone, as far as I can discover, are the windows by which light is let into this dark room: for methinks the understanding is not much unlike a closet wholly shut from light, with only some little opening left, to let in external visible resemblances, or ideas of things without: would the pictures coming into such a dark room but stay there, and lie so orderly as to be found upon occasion, it would very much resemble the understanding of a man, in reference to all objects of sight and the ideas of them (II, 11, 17).

Locke elevates the image of the blank screen to imaginatively portray his view of the understanding as a *tabula rasa,* an empty tablet that needs no innate principles to explain knowledge: 'Let us suppose the mind to be, as we say, white paper void of all characters, without any ideas' (II, 1, 2). Once one enters into the framework of the darkroom root metaphor, Locke's *tabula rasa* supposition seems plausible. On this blank screen internal images appear if the sun lights them sufficiently outside: 'The pictures drawn in our minds are laid in fading colours, and if not sometimes refreshed, vanish and disappear' (II, 10, 5). If our understanding is not refreshed by the sunlit input, then it has nothing to think about, since, he presumes, a blank screen implies no innate ideas. 'The senses at first let in particular ideas and furnish the yet empty cabinet' (I, 2, 15). Locke's argument against innate ideas is pervaded by the built-in presupposition of the darkroom root metaphor. 'Whatever it be that keeps us so much in the dark to ourselves, sure I am that all the light we can let in upon our own minds ... will not only be very pleasant, but bring us great advantage' (I, 1, 1).

Ernest Tuveson (1955) agrees that for Locke the image of the *camera obscura* was a paradigm for contained subjectivity. He stresses that Locke's model of mind, like Descartes's dualism, implies an isolated subjective consciousness that 'cannot stir from its chamber ... cannot look out of the windows of its dark room, that sees only the "pictures" ' (pp. 164–166). Since Locke's representationalism was later adopted as a foundation for much of Western psychology, the *camera obscura* entered into Western thinking as a classic yet largely unacknowledged root metaphor for psyche itself, as if psyche were nothing but a dark room.

THE PRIVATE EYE OF THE SOUL

Before the eighteenth century the *camera obscura* was seized upon as a perfect tool to enhance and express the extreme individualism that nourishes modern subjectivism. In his classic study of Renaissance culture Jacob Burckhardt (1929) outlines several reasons for the rise of individualism by 1400 in Italy. Previously, ancient and medieval self-consciousness was more participatory, for 'man was conscious of himself only as a member of a race, people, party, family or corporation – only through some general category' (vol. 1, p. 143). But the rise of the singular subject followed several developments, so that 'at the close of the thirteenth century Italy began to swarm with individuality' (p. 143). Despotism and rapid changes in political destinies forced many to be aware of themselves as particulars. Banishment, travel, and cosmopolitanism all enhanced individualism. These forces also nourished the rise of literary biographies, artistic portraiture, and humanism (vol. 2, p. 324).

In the midst of these developments the *camera obscura* became a personal, individual experience, a private, contained box for peering out into the world. Early *camerae obscurae* were simply rooms large enough for several people to observe the screen. But soon the darkroom of the soul became an event for one. Small, portable, and even hidden *camerae obscurae* became instruments of private consciousness. Leonardo's note is the earliest surviving description of a portable dark room:

A cubic box should be made of wood with its sides firmly fixed together, except for the one in front which has to be taken over by the plate of iron [with a small hole in it], and the opposite one at the back, which is made of a thin sheet of paper or parchment, glued around the edges of the wooden box (Richter, 1970, vol. 1, p. 133).

This head-like box offered the image of psyche contained in a box. Friedrich Risner was the first to publish the idea of a portable *camera obscura*. In his *Opticae* of 1606, Risner suggested a portable dark chamber for drawing landscapes. Johan Kepler, making a survey in Austria in 1620, used such a portable dark room (Gernsheims, pp. 23–24).

By the mid-seventeenth century, Constantijn Huygens and Athanasius Kircher had each described a portable *camera obscura*. By 1753 the portable *chambre obscure* was common enough to be illustrated in Diderot's *Encyclopedie*, with adjustable external mirrors. Here it was praised as 'an extremely amusing spectacle.' Kaspar Schott described a *camera obscura* from Spain concealed under a cape.

Thus the small portable box camera was common long before the invention of light-sensitive film in 1839. The use of the *camera obscura* rapidly expanded among artists, so that 'in the eighteenth century the camera [*obscura*] was an almost indispensable part of many artists' equipment' (Schwarz, 1949, p. 254). For centuries before the final achievement of photography, the *camera obscura* made available the analogy of psyche as a subjective dark room.

The fantasy of having the psychic power to observe others *secretly* also pressed the *camera obscura* into its service in the seventeenth century. Hidden *chambres obscures* were disguised in many ways. A host could secretly keep an eye on his guests using a 'magic' goblet with a lens in the stem that reflected the view through a mirror set at 45 degrees onto the surface of white wine. The *camera obscura* was also hidden in books, walking sticks, and carriages (Gernsheims, pp. 25–28). Even today in cinematography the dominant custom is that the camera be treated as an unacknowledged, if not secret, 'candid' observer. The candid camera of television is the latest trick in this long line of images of the hidden, subjective psyche.

Portable, secret observation lets the viewer retreat into the little black box of subjectivity and imagine an alien world 'out there' in which she does not participate. It models the fantasy of a purely mechanical world of soulless 'objects' populated with isolated, contained 'subjects' who interact with it by receiving or sending out 'projections.' The *content* of the camera may bind together in equivocal conflict the 'objective portrait' and 'questing subjectivity' that Susan Sontag (1973) has discerned in photography (p. 122). But it seems that the black box *itself portrays* the cognitive map that there *are* two ontologically separate realms: objects peeked at by subjects. Divided at the Private Eye of the lens, the portable, hidden *chambre obscure* separates the world into the external mechanism and the hidden 'ego.'

THE VANISHING POINT OF THE SOUL

The *camera obscura* was one of the often secret tools of Renaissance artists studying the new mental map of perspective drawing. These artists created several devices to help visualize this scheme, as Durer illustrated. Artists who saw the effectiveness of the *camera obscura* as a 'cheat,' suggesting that they could not draw, naturally kept their use of it a secret. But Della Porta let this secret out of the bag in 1558, so that by 1753 portable *camerae obscurae* pictured in Diderot's *Encyclopedia* were commonly known tools for draftsmen. The *camera obscura* was an excellent tool for helping artists project the mental framework of perspective art's geometric worldview. The simplest one-point perspective scheme looks like a pyramid lying sideways, in which the apex is the single point of view for the eye, and the vertical base is the plane for projecting the picture. The vanishing point on the horizon in the picture appears exactly opposite the eye at the apex.

The important philosophical point about this system is that perspective is a system of pictorial conventions laid upon the world like a map. Perspective visualization, whether in drawing or photography, is not the generally assumed accurate picture of nature. As E. H. Gombrich stressed in *Art and Illusion* (1961), 'Perspective is merely a convention and does not represent the world as it looks' (p. 254). These artists's tools were map-makers for the perspective worldview, grids for oversimplifying and distorting the natural world into a new mental construct. Perspective especially aids in the scientific reduction of space to a static, homogeneous, rational framework. As Erwin Panofsky (1924) put it,

> This whole 'central perspective' makes, in order to guarantee the formation of a fully rational, that is to say, endless, stable and homogeneous space, ... a decidedly audacious abstraction of reality (p. 260).

The retina would project in a spherical form, he argues, rather than pyramidal, and photography reinforces this distortion by printing segments of spherical images on flat surfaces. Consequently,

> A fundamental discrepancy is given between the 'reality' and the construction, and also of course the latter is completely analogous to the method of working of photographic apparatus (p. 261).

This reduction of the full reality of space to the map of a merely geometric convention eventually led to the repression of the depth and soul of the world into the unconscious darkroom of the individual, subjective viewer.

Perspective visualization not only distorts the world; it also elevates the individual viewer to the greatest importance. Rudolf Arnheim (1974) emphasizes 'the commonly accepted interpretation of central perspective as a manifestation of Renaissance individualism. The image presents a world as seen from the viewpoint of an individual observer' (p. 294). The single eye of the *camera obscura* is the eye of the subjective psyche that sees things in perspective's design. The dark room is not only a tool used by early creators of perspective art. It is a living concrete experience of the mental map-making exercise that they undertook and imprinted on the Western psyche. The single vanishing point in central perspective is a visual metaphor for the singleness of the individual psyche enshrined as authoritative in this geometric metaphor. And the *camera obscura* is the psyche-box in which this map-making is constantly being undertaken. It is the interiorization and containment of the soul, because the vanishing point is the mirror, not of the infinite universe, but of the sovereign individual, whose point of view determines a vanishing point's position. On the horizon of perspective's program the vanishing point collapses the infinite depth of being into the subjective psyche of an individual with a point of view. 'The world of being,' observes Arnheim, 'is redefined as a process of happening' (p. 298). The vanishing point in perspective envisioning becomes the dividing point on perspective's map between the interior and the exterior worlds. By structuring external space geometrically this imaginary mental map creates an artificial interior world whose function is to be a point of view. The deep mysterious soul of the world is supposedly swallowed up into the little dark room's confining construct. Perspective's little psyche-box becomes the subjective vanishing point of the soul of the world.

THE INTERIOR THEATRE OF THE SOUL

On the boards of the theatrical stage the dramas of the soul are enacted. 'Up there' on the stage the stories of 'in here' are played out, and to the extent that the collective soul of the audience participates in the secret of theatre, the duality of in/out audience/players is mysteriously transcended in a deep collective bond. So, naturally, theatre is a metaphor for the soul. The *camera obscura* inherited this secret and gave it the particular slant of its contained structure, picturing the soul as an internal space. The theatrical darkroom of the psyche became a paradigmatic image of an interior theatre of the soul.

This idea that theatre is a metaphor for the soul was made explicit in the Renaissance memory theatres. As Frances Yates has documented in *The Art of Memory* (1966), an ancient system of memory retention for orators was to store speeches in imaginary theatres. During the sixteenth and seventeenth centuries Giulio Camillo and Robert Fludd developed this system into elaborate Renaissance forms. The trick is to place in memory symbols such as statues in various sequences on a stage. Then to each symbol is attached a concept, so that the speaker's ideas are recalled by a mental walk around the stage. Camillo's stage was organized as the seven pillars of Solomon's House of Wisdom, each standing before a Greek god, such as Apollo, backed up by mythic images such as the banquet and the cave. Fludd's memory stage was based upon Shakespeare's Globe Theatre.

Camillo's and Fludd's memory systems were explicitly religious. Their metaphors for the memory aspect of the soul connected the microcosm to the macrocosm, the human to the ultimate. Camillo actually built a model stage with drawers full of various speeches from classical sources. He sought to store up the eternal nature of all things that can be expressed in speech. These eternal themes were organized under the system of the planetary gods and their ruling passions, such as the tranquillity of Jupiter, the anger of Mars, or the love of Venus (Yates, p. 144). Fludd drew the Globe stage with the circle of the Zodiac on the ceiling (p. 353). In these ways Camillo and Fludd represented the tradition of placing the soul in the context of the cosmos, the individual participating in universal dramas, symbolized by the gods and their stories. The soul's memory was for them no mere interior subjective tool, because it participates in a transcendent drama.

By contrast, during the Renaissance the intellectual fashion known as 'natural magic' began to demystify the many phenomena widely believed to be magic with the new concepts of natural science. While Camillo and Fludd pointed above, an encyclopedist and dramatist named Giambattista della Porta pointed within. He was a sixteenth-century Neapolitan whose widely distributed *Magiae Naturalis* (1558) reductively explained away much fascinating magic, from women's makeup to the *camera obscura*. Della Porta's purpose was to demonstrate that seemingly magical apparitions on the blank screen of the *camera obscura* are merely the consequences of natural principles of optics. Della Porta's account was more complete than any earlier one, and it widely disseminated knowledge of the *camera obscura*. For this reason he was long considered its inventor.

Della Porta's description was enthusiastic, for he believed he was divulging secrets. In 1558 he described

How you may see things in the darkness that are illustrated outside by the sun, and with their colors: Close all the windows. ... The hole may have the form of a cone, with its base facing the sun. Opposite it you arrange a white wall, linen cloth or stretched paper. ... You may see the faces, gestures, movements and dress of men, the dispersements of clouds, the azure sky and flying birds. If you arrive at the truth, you will be greatly pleased (1558, Book IV, chapter 2).

Della Porta also described how the *camera obscura* can be used as an aid to drawing.

If you cannot paint, you can by this arrangement draw with a pencil. Then you will only have to lay on the colors. This is done by reflecting the image downward onto a drawing board with paper (1558, Book IV, Chapter 2).

Here emerges a new task for the *camera obscura*: a memory theatre. But this new memory theatre was a purely mechanical tool. Far from Camillo's and Fludd's picture of memory as microcosmic reflections of the divine macrocosm, Della Porta's *camera obscura* for artists was merely a mechanical memory tool. While Camillo and Fludd placed the soul's *memoria* in the cosmic drama, Della Porta deposited memory in the mechanical box that would become the interiorized photographic record of the neutral, objective construct of Camera Man.

But Della Porta was not content with using the darkroom as a memory instrument. He was also a playwright, so he cleverly arranged theatrical productions with scenery, actors, and music on a sunlit stage outside the darkroom to surprise guests inside. Satisfied with his trick, he wrote, 'In a chamber you may see hunting, battles of enemies and other delusions. ... Nothing can be more pleasant for great men, scholars and ingenious men to behold' (1589 ed., Book XVII, chapter 6). But spectators in the dark room were so alarmed by his projected spectacle that they got him into trouble. Della Porta must have wanted to perform magic more than expose it, because his guests could not believe his explanation. They accused him of sorcery.

This earliest forerunner of the cinema signaled a crucial turn in the development of the latent fantasy of the dark room as the psyche. No longer used only for optical experiments, or simply as a recording device for the memory, the *camera obscura* now became a means of intentionally projecting fantasy. The psychic functions of the darkroom now embrace observation, memory and imagination. Early viewers of these enchanting projections were deluded. They could not imagine the possibility of naturally projected images and so they attributed Della Porta's trickery to sorcery. The psychic activity in the dark room was now dramatically intensified. In the *camera obscura* fearful delusions were generated, for psyche was stirred in a new way, and souls were moved to terror and wonder. In Della Porta's dark room, psyche appeared alive as never before. To viewers today, who have become so blasé about cinematic techniques, the terror and the wonder have become commonplace in cinematic spectacles. But to Renaissance audiences the experience of all this soul activity in a dark room was electrifying. The *camera obscura* becomes the interior theatre of the soul. The stage becomes the white screen and the window to the gods becomes the flickering projections of the subjective psyche.

THE PROJECTED METAPHYSICS OF THE SOUL

The *camera obscura*, we have seen, has helped structure the subjectivist model of the psyche as an observing instrument (Leonardo), as a blank screen (Locke), as a private individual container (Diderot), as a perspective-designing tool (Arnheim), and as an interior theatre (Della Porta). When it came to religion, there was no shortage of applications. The characters of metaphysical mythology were rapidly consigned to the skull's darkroom.

During the Renaissance, the ghosts, devils, and demons of popular religious mythology, long believed to influence events from dark mysterious places, began to appear moving about in the new psyche-box. Itinerant performers used the *camera obscura* to frighten audiences in dark fairground booths. A detailed exposé of their techniques was recorded in Belgium by Francois D'Aguillon in 1613:

In just this way certain charlatans tend to hoodwink the uneducated rabble; they pretend to know about black magic, while they are hardly aware of what that means. They boast that they conjure up phantoms of the devil from hell itself and show them to the onlookers. They lead the inquisitive and curious, who want to know all about secret and obscure subjects, into a dark chamber where there is no light, except a little which filters through a small pane of glass (lens). Then they tell them sharply not to make any noise and to be as quiet as a mouse. And when everything is completely still and no one either moves or says a word, as if they were waiting for a church service or a vision, they say that the devil will soon come. In the meantime, an assistant puts on a devil mask to make him look like the pictures of devils one usually sees, with a hideous, monstrous face and horns on the head, with wolf's pelt and tail, with claws on his sleeves and shoes.

Then the assistant struts up and down outside (the camera) as if he were deep in thought, to the place where his colours and shape can be reflected through the glass pane into the chamber. To make these cunning intentions even more effective, one should remain absolutely still, as if a God were to arise by way of this artifice. Then a few begin to go pale; some, out of fear of what is to come, start to sweat. After that, one takes a large piece of cardboard and stands it opposite the light-rays which have been allowed to enter the chamber. On it can be seen the picture of the devil walking up and down; this they look at with trepidation. This is the reason why the poor and unexperienced are unaware that they see the charlatan's shadow and squander their money unnecessarily (Hecht, 1984, p. 5).

The devil, commonly pictured in sermons, church art, and popular theatre, now dared to appear moving about mysteriously in the *camera obscura*. While earlier static images were understood as conventions, the new, strangely moving images inside these dark rooms were terrifying. But much as the fear of thunder and lightning is diminished by scientific explanations, fear of the devil can be defensively decreased by reducing him to natural magic's trickery. The person who stood up in that old fairground booth and held the cardboard in front of the incoming projection, to reveal the artifice, was helping to reduce visionary metaphysics to natural optics.

Exposés of shows like this eventually supported the argument that mythic creatures such as the devil are 'nothing but projections.' Once the actor outside the room is unveiled, the psychic event inside the room seems less fearful. Such demystification of imaginative figures implied the correlative assurance that, 'It's all in your head.' The terror, the guilt, the evil evoked by the devil can now be imagined as merely the passions felt inside the darkroom of the soul. The emotion experienced in the *camera obscura* is relieved upon discovering the 'objective facts' outside (the actor outside disguised as the devil). That relief can turn into iconoclasm, skepticism, even a defense against the soul's passions. The contrast between 'external fact' and 'internal feeling' is then more plausible, and is even called upon to support a dualistic metaphysics. In this scheme the darkroom of the soul becomes the basement of reason's subjective castle, the dungeon for all the formerly literalized, external metaphysical images of religious dogma. Slowly but surely, the inner self and interior feelings develop into the new locus for devils and angels alike. Eventually anyone believing in angels, devils, or God was accused of foolishly 'projecting' them out of their 'subjective' soul into the 'objective' world. The magic lantern, recently renamed the slide projector, aided in this conceptualization, as I have shown elsewhere (Bailey, 1986).

By the nineteenth century the projection metaphor became a serious critique of religion. When Ludwig Feuerbach argued in 1841 that 'what by an earlier religion was regarded as objective is now regarded as subjective' (p. 13), he presumed not only the subject/object dichotomy, but also the hidden image of the *camera obscura*. The real experience of gods and devils occurs only inside the subjective dark room of the soul, he proposes, and is only erroneously externalized or projected outside it.

Similarly, when Sigmund Freud (see 1974) later postulated that most mythology and religion 'is nothing but psychology projected into the external world,' he imagined psyche as a contained, internal reservoir erroneously flowing out into the Newtonian machine-world (vol. 6, p. 259). Mythology belongs inside the head, he presumed. Like the dark booth at a Renaissance fair, psyche for Freud belongs inside the container for soul's emotions, and anything mythic seen outside is like a magician's trick. The only reality of soul's experiences lies inside the psyche-box, the root metaphor teaches.

THAT OBSCURE CHAMBER OF THE SUBJECT

This idea that the soul is restricted to an internal chamber, fundamental to Western thought since Descartes, is an unnecessarily restrictive distortion of psyche's natural participation in the world. This subjectivism, reinforced by the experience of the *camera obscura*, serves the Newtonian notion that the world is like a vast collection of objects, organized into a machine, which has no place for psyche except inside our dark skulls. To serve physics, subjectivism scrubbed the world clean of participation in culture, nature, and being, clearing the way for the development of technology. But this scrubbing represses our awareness of the extent of our participation. And this repression has become seriously problematic.

As the problems of the mechanistic worldview have become evident, ecologists have warned us that we urgently need to realize our interconnected relations with the whole of nature. Theoretical physicists have told us that, 'Quantum theory forces us to see the universe not as a collection of physical objects, but rather as a complicated web of relations between the various parts of a unified whole' (Capra, 1975, p. 124). Consequently, a network of various approaches is developing whose common theme can be termed 'participation.' Morris Berman sees 'participating consciousness' as 'the next step in the creation of a 'post-Cartesian paradigm,' in his *The Reenchantment of the World* (1981, p. 152). He builds on Owen Barfield's distinction between a naive, superstitious 'original participation' and a post-modern, sophisticated 'final participation' (*Saving the Appearances*, 1957). Berman also points to Gregory Bateson, whose holistic systems studies have opened new paths to understanding participation. James Hillman's (1975) approach to archetypal psychology has further opened such creative vistas by exploring images of soul-in-the-world. And important strands of thought such as the phenomenology of religion increasingly echo Heidegger's fundamental opening to being-in-the-world. Rethinking the primordial experience of participation moves us out of the *camera obscura*'s fantasy of subjectivism.

But the image of the *camera obscura* continues to unconsciously suggest that psyche still inhabits a dark room illuminated by inwardly projected images. Unless we become conscious of this influence, the root metaphor of the dark room containing psyche will continue to secretly distort our concept of soul. As a paradigm for isolating subjectivism, the *camera obscura* needs to be demystified so the imprisoned soul can break out of its long repression in the dark dungeon of skull's darkroom.

Ithaca College

REFERENCES

Aristotle. *On Memory and Recollection* (450a).
Arnheim, Rudolf. *Art and Visual Perception* (Berkeley: University of California Press, 1974).
Bailey, Lee W. 'Skull's Lantern: Psychological Projection and the Magic Lantern,' *Spring 1986*, pp. 72–87.
Bacon, Roger. *Perspectiva*. (Frankfurt, 1614).
Barfield, Owen. *Saving the Appearances: A Study in Idolatry* (New York: Harcourt, Brace and World, 1957).
Barfield, Owen. *The Rediscovery of Meaning, and Other Essays* (Middletown, Conn.: Wesleyan University Press, 1977).
Berman, Morris. *The Reenchantment of the World* (Ithaca: Cornell University Press, 1981).
Burckhardt, Jacob. *The Civilization of the Renaisance,* 2 volumes (New York: Harper and Row, 1929).
Capra, Fritjof. *The Tao of Physics* (New York: Bantam, 1975).
Della Porta, Giambattista. *Magiae Naturalis* (Antwerp, 1558). This book went through 23 editions in Latin, and numerous translations. First published in 1558 in four sections, it was expanded to twenty in the 1589 edition. In the 1558 edition, the *camera obscura* is discussed in Book IV, chap. 2, and in the 1589 edition, in Book XVII, chap. 6. An English translation of the later edition was published in London in 1658.
Descartes, René. 'La Dioptrique' (1637). Discourse 5 in *Oeuvres de Descartes* (Paris: Vrin, 1913).
Feuerbach, Ludwig. *The Essence of Christianity* (1841), trans. George Eliot (New York: Harper and Row, 1957).
Freud, Sigmund. *The Standard Edition of the Complete Psychological Works of Sigmund Freud,* ed. J. Strachey (London: Hogarth, 1974).
Gernsheim, Helmut, and Alison Gernsheim. 'The History of the *Camera Obscura*,' in their *The History of Photography* (New York: McGraw-Hill, 1969), pp. 17–29.
Gombrich, E. H. *Art and Illusion* (New York: Pantheon/Bollingen, 1961).
Harris, John. 'Camera obscura,' in *Lexicon Technicum* (London, 1704).
Hecht, Hermann. 'The History of Projecting Phantoms, Ghosts and Apparitions,' *The New Magic Lantern Journal* 3 (February, 1984): 2–6.

Hillman, James. *Re-Visioning Psychology* (New York: Harper and Row, 1975).
Huygens, Constantijn. *De Briefwisseling* (1608–1687), ed. J. A. Worp (The Hague: Nijhoff, 1911). See Arthur Wheelock, 'Constantijn Huygens and Early Attitudes toward the *Camera Obscura,*' *History of Photography* 1 (April, 1977): 93–103.
Kofman, Sarah. *Camera Obscura de l'idéologie* (Paris: Editions Galilée, 1973).
Locke, John. *Essay on Human Understanding* (New York: Dutton, 1976).
Panofsky, Erwin. 'Die Perspektive als "Symbolische Form,"' *Vorträge der Bibliothek Warburg* (1924), pp. 258–354.
Pepper, Stephen. *World Hypotheses* (Berkeley: University of California Press, 1942).
Richter, Jean. *The Literary Works of Leonardo,* 2 volumes (Berkeley: University of California Press, 1970).
Rorty, Richard. *Philosophy and the Mirror of Nature* (Princeton: Princeton University Press, 1979).
Ryle, Gilbert. *The Concept of Mind* (New York: Barnes and Noble, 1949).
Sabra, A. I. 'Ibn al-Haytham,' in *Dictionary of Scientific Biography* (1970), vol. 6, pp. 189–210.
Schwarz, Heinrich. 'Art and Photography: Forerunners and Influences,' *Magazine of Art* 42 (November, 1949): 252–257.
Sontag, Susan. *On Photography* (New York: Delta, 1973).
Tuveson, Ernest. 'Locke and the "Dissolution of the Ego,"' *Modern Philology* 52 (1955): 159–174.
Yates, Frances. *The Art of Memory* (Chicago: University of Chicago Press 1966).

EDMUND F. BYRNE

WORKPLACE DEMOCRACY FOR TEACHERS: JOHN DEWEY'S CONTRIBUTION

John Dewey may prove to have been one of America's most important philosophers of technology. Like Thorstein Veblen, however, Dewey took technology to encompass all applications of scientific method to the solution of human problems. Thus his 'instrumentalist' approach to problem-solving included, indeed came to focus on, social organization as a *sine qua non* for enhancing the democratic ideal. Especially important in this respect was Dewey's persistent dedication to the well-being of working people. He was, to be sure, an intellectual but a distinctively pro-labor intellectual. More than that, he was an activist in behalf of unions – for 'handworkers' and 'brain-workers' alike. In this cause he had few allies among academicians. Now, however, three decades after his death, American professors are beginning to unionize – hopefully with the same breadth of vision that motivated Dewey.

This offbeat encomium of the great philosopher will no doubt please few Dewey scholars, but for quite different reasons. Leftists like Garry Brodsky, in keeping with their forebears' approach during Dewey's lifetime, dismiss this concern for workers as empty bourgeois rhetoric. But they may be reacting less to Dewey than to his right-wing interpreters, such as Sidney Hook, Ernest Nagel, and Lewis Feuer. Both sides miss the point just to the extent that they read Dewey too narrowly through their own lenses, neglecting his lifelong search for justice in a world marked by accelerating concentration of power and wealth. Conservative defenders of Dewey, in particular, fail to include in their interpretation of his views any account of his pro-union orientation. The resulting caricature borders on being schizophrenic.

Lewis Feuer, for example, says Dewey was an inept experimentalist; but because of his 'scientific character' he was our 'greatest intellectual.'[1] This title he awards to Dewey because he helped Leon Trotsky modify his belief in the socialist system; belatedly spotted the treachery of the Soviets while other intellectuals were still soft on Communism; and predicted the Nazi-Soviet Pact. He does not honor Dewey for supporting low-vote-getter Socialist Norman Thomas, and certainly not for his longtime heroic support of the rights of workers in the U.S. Instead, Feuer calls this an emotive attachment, uses the Soviet system as a slippery slope objection, and asks if Dewey has not perhaps exceeded 'the experimental constraints that an instrumentalist pragmatism might have placed on his socialist enthusiasm.' Having thus set the trap, Feuer springs it on Dewey by using, he thinks, Dewey's own method against him:

> Dewey's faith that ordinary citizens can be relied upon to make rational decisions with respect to basic political issues is itself a hypothesis to be verified, limited, or contravened; and when Walter Lippman challenged it, Dewey actually adduced little evidence against him.[2]

This Hamiltonian skepticism about democracy Dewey had in fact countered (a) constructively, by his lifelong commitment to education of the working class, and (b) explicitly, in his 1928 article, 'Philosophies of Freedom':

> Practically, every one admits that there is a new social problem, one that everywhere affects the issues of politics and laws; and that this problem, whether we call it the relation of capital to labor, or individualism versus socialism, or the emancipation of wage-earners, has an economic basis. The facts here are sufficient evidence that the ideals and hopes of the earlier liberal school have been frustrated by events; the universal emancipation and the universal harmony of interests they assumed are flagrantly contradicted by the course of events. ...[3]

To explain how this 'earlier liberal school' could have been so wrong, Dewey points to what he calls 'the real fallacy.' This, he says,

> lies in the notion that individuals have such a native or original endowment of rights, powers, and wants that all that is required on the side of institutions and laws is to eliminate the obstructions they offer to the 'free play of the natural equipment of individuals.' ... The notion that men are equally free to act if only the same legal arrangements apply equally to all – irrespective of differences in education, in command of capital, and the control of the social environment which is furnished by the institution of property – is a pure absurdity, as facts have demonstrated. ... The only possible conclusion, both intellectually and practically, is that the attainment of freedom conceived as power to act in accordance with choice depends upon positive and constructive changes in social arrangements.[4]

In some respects Dewey may well have been our 'greatest intellectual,' as Feuer contends. But, I submit, there are better reasons for assigning him such an honorific title than because of some passing success in applying the scientific method to a social problem. In particular, he merits comparable recognition because of his lifelong battle for workplace democracy despite lack of support from the American intellectual community.

I. DEWEY'S ACTIVIST DEFENSE OF UNIONS

Dewey's academic career, which began in 1884 at the University of Michigan and ended in 1939 at Columbia University, spanned the crucial years during which American workers struggled for and won the basic rights to organize and bargain collectively the terms and conditions of their employment. Throughout that period he proved himself a staunch defender of the labor movement.

His lifelong commitment to industrial democracy was already in evidence in the earliest years. All over the country, workers to whom the law granted few rights were challenging the autocracy of capitalist entrepreneurs; and their challenge was frequently countered by systematic, government authorized, if not initiated, violence, the most notorious instance of which occurred at Haymarket Square in Chicago in 1886. At this time Dewey was teaching an ethics course on the 'disequilibration' of capital and labor; and in 1888 he published *The Ethics of Democracy*, in which he argues that economic and industrial democracy is necessary to avoid the kind of violence that was then so tragically characteristic of industrial relations. Later, during World War I, he joined with Thorstein Veblen and others in actively protesting government persecution of the controversial International Workers of the World.

During his long tenure at Columbia University, Dewey matched his well-known attention to education with comparably dedicated attention to the concerns of educators. He organized and actively participated in unions, pioneered in labor education, and was a front-line activist in behalf of rights he considered essential to industrial democracy.

In the years immediately before World War I, he was a key factor in the organization of the New York Teachers Union and its affiliation with the American Federation of Teachers, which was itself affiliated with the American Federation of Labor, as Local No. 5. He served for three years as its first vice-president. Over the years he repeatedly defended the propriety of teachers' belonging to unions in public addresses which in published form influenced ever wider audiences.[5] In the 1930s, when over seventy years of age, after failing to prevent a Communist takeover of his Local No. 5, he led 900 others from that union to establish the New York Teachers Guild; and he provided similar support to embattled unions elsewhere: AFT Local 195 at Cambridge, Massachusetts; the National Education Association (1931); and the New Haven Teachers Association (1933).[6]

Education was for Dewey the key to developing a working class that would make democracy workable. Over the years, he showed his commitment to *labor* education in many ways, e.g., obliquely, by opposing ROTC programs in colleges and universities and, more obviously, by opposing the proposal of certain manufacturers to put academically unpromising students over fourteen to work in factories (rather than providing alternative learning).[7] The New School for Social Research, of which he was a co-founder, was envisioned as a center of learning where political, social, and economic problems could be studied in a free atmosphere, without fear of recrimination. But the focal point of Dewey's commitment to labor education was the Brookwood Labor College at Katonah (Westchester County), New York. Growing out of the socialist-oriented 'X' Club (organized in 1903), Brookwood was financed by leading social progressives and various labor unions with the aims of applying Dewey's educational philosophy to industrial relations. Increasingly it was branded as socialist by its right-wing opponents, including A. F. of L. vice president Matthew Woll. Reacting to Dewey's favorable report on the Soviet Union, Woll denounced Dewey as 'a propagandist for Communist interests' at the union's 1928 convention; this led to the AFT's withdrawal of support for the Labor College.[8] Dewey complained that this condemnation was based on no investigation whatsoever and was mainly a ploy on the part of Woll to further his goal

> to eliminate from the labor movement the schools and influences that endeavor to develop independent leaders of organized labor who are interested in a less passive and a more social policy from that now carried on by the American Federation of Labor. ...[9]

These words did not save the Labor College. It had to be closed for lack of funds. But they do reveal Dewey's objectives, which he pursued more broadly in the public arena.

In addition to his efforts in behalf of teachers' unions, John Dewey helped organize and direct a number of organizations whose objectives included better public understanding of the legitimate demands of the working class. He was singularly responsible, along with fellow philosopher Arthur O. Lovejoy, for the founding of the American Association of University Professors (AAUP), served as its first president (1915), remained active in the organization and was eventually awarded an honorary life membership. He was one of the founders of the American Civil Liberties Union (ACLU) in 1920, and served on its National Committee through that decade. Dewey joined the League for Industrial Democracy in 1925; and he was the first president of the League for Independent Political Action (LIPA), which was organized in 1930 to reform New York City government.[10] From 1929 to 1936 he served as first president of the pro-labor People's Lobby and wrote numerous articles in its bulletin that called for social protection of the unemployed.[11]

Dewey was also an outspoken defender of the rights of academicians, which he based not on 'academic freedom' but on the importance of protecting the university's responsibility to the public rather than to its administration. Thus did he defend (1) a pro-labor economist fired in 1915 by the University of Pennsylvania; (2) three anti-war professors dismissed in 1917 by Columbia University (Dewey himself supported America's involvement in World War I); and (3) philosopher Bertrand Russell, whose chair in philosophy the City University of New York took from him in 1941 because of public disapproval of his views about sexual morality.[12]

II. DEWEY'S THEORETICAL DEFENSE OF UNIONS

The rights that Dewey sought to secure for workers through his actions he endeavored to justify in his writings, especially by appealing to instrumentalist principles. These principles, in turn, were themselves a function of the social, political, and economic problems that dominated the American scene throughout his lifetime. I have already noted his early commitment to industrial democracy. This objective he explained in 1916 by saying that the growth of democracy is connected to the development of (1) the experimental method in the sciences, (2) evolutionary ideas in the biological sciences, and (3) industrial reorganization.[13] The latter necessitates reform; evolutionary ideas suggest that reform is possible; and the experimental method provides a non-violent means to bring it about. Thus twenty years later Dewey was advocating socialism and freedom of assembly in the workplace as ideas that flow necessarily from these democracy-building developments.

Writing in 1938, Dewey pointed out that ordinary people can use new technologies very well once they are introduced into 'the organized means of associated living.'[14] These organized means of associated living also include, for Dewey, the results of the activity he variously called 'thinking,' 'the method of intelligence,' 'the method of inquiry,' and, most descriptively of all, 'the method of cooperative experimental intelligence.'[15] This social technology involves 'the intervention of inquiry in the way of observation, inference and reasoning.'[16] Inquiry does not take place in isolated minds, as was imagined by the atomistic theory of the early liberals, but requires the collective interaction of the inquirers. Thus subjected to social control, inquiry would be far more reliable as a guide to public policy than is the propaganda disseminated by politicians and pressure groups.[17]

The awesome agenda of Dewey's social technology was already in place in 1916, when he recommended 'a thoroughgoing and constant dependence upon the practice of science' in such enterprises as diplomacy, politics, and ethics.[18] If only philosopher Charles Sanders Peirce's 'laboratory habit of mind' could be carried into such fields as politics and law and political economy, intellectual changes beyond prediction would come about.[19]

In the 1930s, Dewey's democratic agenda remained essentially unchanged; but by then he had learned to be more guarded about the ease with which it might be carried out. Social phenomena, he now suspected, could not be understood independently of 'extensive prior knowledge of physical phenomena and their laws.'[20] Yet he was now persuaded that effectuation of the agenda had become critically important. The 'new force generated by science and technology,' he had to admit, 'has not accrued to the betterment of the common human estate' in anything like the degree that Francis Bacon had anticipated. The institutional framework of society was built upon 'coercion and oppression on a large scale.' This failing, however, was due not to scientific method and technology but to 'institutions and habits originating in the pre-scientific and pre-technological age' and as yet 'untouched by scientific method.' These obsolete institutions and habits society must discard – not by resorting to violence in the name of the class struggle, especially in view of the destructiveness of technology-based modern warfare, but through 'the impact of inventive and constructive intelligence.'[21]

At stake, according to Dewey, was the achievement of a 'vital and courageous democratic liberalism': 'the one force that can avoid reducing the issue for the future to a struggle between Fascism and Communism.' This collectivistic liberalism, unlike its timid predecessor, endorses 'organized social control of economic forces.'[22] In particular, the laissez-faire liberal who objects to a publicly funded social security system in the name of 'rugged individualism' fails to recognize that

servility and regimentation are the result of control by the few of access to means of productive labor on the part of the many.[23]

This being the case,

regimentation of material and mechanical forces is the only way by which the mass of individuals can be released from regimentation and consequent suppression of their cultural possibilities. ...[24]

Stated still more strongly,

> The only form of enduring social organization that is now possible is one in which the new forces of productivity are cooperatively controlled and used in the interest of the effective liberty and the cultural development of the individuals that constitute society.[25]

The context of these assertions is, of course, the Great Depression. But Dewey's commitment to the democratic potential of working people was in place much earlier. It was at the heart of his attack on Taylorism in 1916. Recalling Plato's definition of a slave as 'one who accepts from another the purposes which control his conduct,' Dewey warned that such slavery would surely result from the so-called 'scientific management of work.' Rejecting this 'narrow view' of the role of science with regard to workers, Dewey called upon scientists to address the 'technical, intellectual and social relationships involved in what [workers] do.'[26]

For Dewey, then, the whole person goes not only to school but also to work. And it is precisely this consideration that constitutes one of Dewey's principal objections to laissez-faire liberalism, namely,

> that it conceives of initiative, vigor, independence exclusively in terms of their least significant manifestation. They are limited to exercise in the economic area. The meaning of their exercise in connection with the cultural resources of civilization in such matters as companionship, science and art, is all but ignored.[27]

Thus is American collectivistic liberalism set apart from British individualistic liberalism by its openness to 'the use of governmental action for aid to those at economic disadvantage and for alleviation of their conditions.'[28] This involvement of government he defends along modified libertarian lines, asserting that the police function of the state encompasses protection of our constitutionally guaranteed freedom of association. Writing just at the time when Congress was debating the future National Labor Relations Act (1935), Dewey says:

> It is the business of the state to protect all forms and to promote all modes of human association in which the moral claims of the members of society are embodied and which serve as the means of voluntary self-realization.[29]

Such government protection of the right to organize is of crucial importance because, in spite of pre-industrial individualist values, 'the isolated individual is well-nigh helpless' in the modern industrial setting where 'concentration and corporate organization are the rule.' Concern about giving the state too much power is misplaced, since '[the state's] exercise of coercive power is pale in contrast with that exercised by concentrated and organized property interests.'[30]

This countervailing-force defense of state intervention to protect working people's interests is clearly not statist in intent. For Dewey's confidence in the potential of face-to-face democratic process remained the very touchstone of his *praxis,* setting his position apart, at times rather subtly, from the more radical responses which he rejected. What is here manifested, then, is an act of desperation on the part of a dedicated guild socialist who has come to recognize that democratic values would not survive without government protection in behalf of the people.

As Arthur Lothstein has noted, Dewey does not provide us with a full-blown description of how worker self-management might be carried out.[31] But neither has he simply left us in the dark on this subject. For his own activism provides a number of relevant and historically significant examples of professional interest-group self-management, not in isolation from but in responsible attention to the public good. 'Ideas' suitable for incorporation into public policy are to be developed co-empirically, by the method of intelligence. Or, as he put it in his later years, a liberal program has to be developed independently of government and 'enforced upon public attention, before direct political action of a thoroughgoing sort will follow.'[32] To the methodology for developing that liberal program Dewey devoted much of his attention.

Showing in his earlier years the influence of Thorstein Veblen, Dewey spoke of applying the methodology of scientists and engineers to bring about 'the invention and projection of far-reaching social plans.'[33] In later formulations he tended more to an account that maintained an epistemological equivalence between comparably reasoned conclusions about quite different kinds of subject matter. As he put it in *The Quest for Certainty* (1929), one may arrive at unqualifiedly scientific conclusions about 'fuller and more complex social and moral affairs' if one proceeds methodically, method being the determinant of a cognitive conclusion's value.[34] Epistemological subtleties aside, reasoning, as Dewey understood it, articulates a problem from experience and projects the possible consequences of a tentative solution. If this solution is implemented in action, it is the task of thinking to discover the specific connections between the action taken and its actual consequences. In this way 'inquiry' transforms an indeterminate situation into a unified, partially determinate whole.[35]

This, Dewey repeatedly warned, cannot be accomplished by any mechanistic application of abstract science to the world of common sense. That kind of so-called 'applied science,' which reduces qualitative matters to the quantitative and final to efficient causes, is responsible for a great deal of social disintegration; and this, says Dewey, is why people make such a sharp distinction between 'common sense inquiry and its logic and scientific inquiry and its logic.'[36] If democracy is to be truly viable, the method of inquiry must be used not as an external technology that mechanizes life and enslaves people but rather in a 'naturalized' (i.e., interest group-originated) way.[37] Only so can *the ancient social dichotomy between free citizens and workers* at last be abolished. And once this social dichotomy has been abolished, *the epistemological distinction between theory and practice, which arose out of that dichotomy,* will collapse.[38]

This, in briefest outline, is how Dewey attempted to justify philosophically the right of workers to organize – a right which, as we have seen, he himself defended actively over the course of his life as a dedicated unionist. In fact, his activism increased in his later years. Yet even at the time of Dewey's death, in 1952, very few American university professors belonged to a union. Now many do. What changes have brought this about? In particular, has the American professoriate as a group come to share some of the political insights that motivated Dewey's unionism?

III. AMERICAN PROFESSORS IN UNIONS

American professors during Dewey's lifetime thought of themselves as elite professionals, a view that Dewey shared. This collective self-image changed little during the first twenty years after his death. The post-war expansion of American higher education brought many middle and even working class people into the ranks of the professoriate. But so long as the 'sellers' market" for qualified academicians prevailed, these new arrivals were able to disregard the Depression-era concerns of their forebears. They moved freely, like independent contractors, from one academic job opportunity to another. By 1972, however, the euphoria of the 'brick-and-mortar' era had dissipated. Their job opportunities having dwindled, academics had to adjust to the less autonomous role of comparatively immobile employees. Like other professionals, they too began to share, in ever increasing numbers, John Dewey's recognition of the benefits of organization. In fact, since 1972 American professors, along with other white collar workers, have constituted the fastest growing sector of the unionized workforce. This coming to terms with reality amounts, however, to a reluctant accommodation to facts that have yet to be internalized. What appears to be lacking in this will-to-unionize is precisely the sort of theoretically grounded political sensitivity that Dewey brought to his role as an academician. Consider first the basic data in this regard, then some interpretive observations.

The unionized segment of the total workforce in the United States fluctuated slightly around an average of twenty-three per cent between 1973 and 1981, according to U.S. Department of Labor statistics. This stability is not what one would expect in view of the high unemployment and worker displacement during that period of time. For, due to these 'hard times,' there was a sharp decline in union membership over a wide range of industries, especially in manufacturing. Fortunately for the figures, blue collar attrition was counterbalanced by the rapid increase in unionization among white collar workers, including college and university faculty.

In 1970 about 5 million white collar workers belonged to unions; in 1980, almost 8.5 million, representing 37.8% of all unionized workers. And of this total almost half, i.e., about 4 million, were professional and technical workers. Overall union membership among professional, technical, and kindred workers in the United States rose from just under 14% in 1974 to nearly 30% in 1980. And the two sectors most responsible for this increase are health care and especially education.

The extraordinary transformation in the role of health care professionals, including physicians, is due to many factors; but chief among these is the intensification of government regulation coupled with a decline in government funding and the correlative privatization of health care facilities. In this context the percentage of unionized registered nurses rose from barely 8% in 1974 to 16.5% in 1980. Comparable increases occurred among pharmacists, chiropractors, health technologists/technicians, radiologic technologists/ technicians, and even health administrators. Among related helping professionals there were some even more dramatic increases: psychologists (from about 9% to 22%), social workers (from under 20% to 29%), and social welfare clerical assistants (from 17.5% to 43%). A comparatively smaller number of physicians have joined unions; but considering the income level of this profession (average annual income over $100,000), it is remarkable that 25,000 out of a total of just over 500,000 doctors had organized by 1985.[39]

Similar and in some respects even more impressive increases have occurred in John Dewey's own bailiwick, that of education. This is especially true of the K-12 sector; but, as already noted, there have been noteworthy increases in higher education as well.

The level of organization among different groups of teaching professionals is discernible from U.S. Department of Labor statistics. Between 1974 and

1980 the percentage of unionized teachers on the preschool and kindergarten level rose from under 12% to 24%; unionized adult education teachers, from 14% to 22%; unionized elementary school teachers, from 26% to 54%; and secondary school teachers, from under 29% to nearly 60%; and trade and industrial teachers, from none to 70%.

The overall impact on different industries served by professionals can be summarized as follows. From 1974 to 1980 the percentage of the workforce that was unionized in welfare services rose from under 17% to over 21%; in libraries, from 9% to almost 20%; in hospitals, from 13.4% to 17.6%; and in elementary and secondary schools, from, roughly, 22% to 42%. Considering these increases in terms of governmental unit, local government workers were under 27% unionized in 1974, over 37% in 1980; state government workers rose from under 19% to over 29%; and federal government workers, up from under 16% to over 18%. During this period the percentage of union members among all college and university employees doubled, going from under 9% to over 16%.

Meanwhile, there has been a significant increase in the percentage of college and university teachers who are unionized. In fact, unionization in this sector is occurring at a more rapid rate than in the workforce as a whole. This increase can be documented by comparing (a) figures reported in 1974, (b) relevant figures in the 1974–1980 Bureau of Labor Statistics report just considered, and (c) figures reported in 1984 by the National Center for the Study of Collective Bargaining in Higher Education and the Professions.

Its members having decided in 1972 to compete with the NEA and the AFT in bargaining elections, the previously aloof American Association of University Professors in 1972 published a *Primer on Collective Bargaining for College & University Faculty*. In the introduction a list of higher education institutions with faculty bargaining agents is preceded by the following statement:

From the first four-year institution to select a collective representative in 1968, there are now more than 71 four-year colleges and universities, private and public, whose faculties are collectively represented. This includes multi-campus state systems in New York, Pennsylvania and Hawaii, as well as small private colleges such as Bard in New York and Regis in Colorado. In addition, the faculties of more than 240 two-year institutions are engaged in collective bargaining. As many as 80,000 faculty members, representing perhaps 15 percent of the American professoriate are now collectively represented.[40]

Job categories used by the Bureau of Labor Statistics do not neatly distinguish college and university faculty from their K-12 colleagues or from comparably trained professionals who are not academics. But the Bureau reports that 'Miscellaneous Teachers, College & University' were unionized barely 1% in 1974 and over 20% in 1980; and 'Teachers, College/University, not specified' rose from under 7% in unions in 1974 to over 13% in 1980.

By 1984, according to the National Center for the Study of Collective Bargaining in Higher Education and the Professions,

about 25% of the 3,200 colleges and universities in the United States had unionized faculties. As of that year there were 429 recognized bargaining agents, of which 355 were at public institutions.

These figures are remarkable for several reasons. In the nation as a whole, the unionized sector of the workforce dropped, according to the Bureau of Labor Statistics, from 23% to 18%. Second, the U.S. Supreme Court's 1980 *Yeshiva University* ruling that private university faculty are managerial personnel and hence not covered under the NLRA has effectively put a lid on organizing in that sector. In fact, before *Yeshiva* more than 90 private institutions were unionized, and by the end of 1984 that number had dropped to 62. Third, the rate of increase in the number of states with enabling legislation for public higher education bargaining slowed almost to a halt during this period, with Illinois one of few states to be added to the list in 1983.[41] In view of these counterindications, the manifestly established trend towards collective bargaining in higher education requires some explanation.

In 1973 the Carnegie Commission on Higher Education published a study by Everett Carll Ladd, Jr., and Seymour Martin Lipset entitled *Professors, Unions, and American Higher Education*. According to Ladd and Lipset, almost two-thirds of faculty surveyed in 1969 favored the idea of collective bargaining and about half the idea of going on strike; and 'after three years in which unionization had made considerable headway, academe was evenly divided as to its benefits.'[42] In their opinion,

the rapid growth of collective bargaining in higher education during the past half-decade should be seen as the extension – to the level of university governance and faculty life – of the powerful trends toward equalization and away from elitism that have characterized many sectors of American society since the mid-sixties.[43]

While an interest in expanding democracy is not to be discounted, less abstract explanations are ready to hand, and are not all that different from the reasons for bargaining increases among professionals outside of academe: concern about the decline of autonomy and professionalism as a result of societal intrusions of all sorts. In addition to these factors, there may be added as catalysts for faculty unionization an impoverishing salary level that, combined with other demoralizing working conditions, threatens to drive as much as forty percent of the American professoriate, especially those in the arts and sciences, out of academe within the next five years. This situation is by no means unique to the United States; but it is not for that reason any less serious. According to one analyst of the trend, Michael I. Sovern, the resulting impact on higher education is 'a potential disaster for the country and for the chain of human knowledge.'[44]

Faculty engagement in collective bargaining is hardly a panacea for all that ails our colleges and universities. But it can counterbalance the many forces indifferent to the value of knowledge that are infiltrating and undermining what we still call higher education. If a faculty union is to function as a defender of intellectual integrity, however, its members must be motivated not only by the bread-and-butter incentives to which nationally affiliated organizers typically appeal but also and indeed preeminently by the direct responsibility of 'brainworkers' to the public.

Happily, we may more easily attain this broader perspective if we stand on the shoulders of John Dewey – not to see merely what he saw but to know better how to look.

Indiana University, Indianapolis

NOTES

[1] Lewis S. Feuer, 'Was John Dewey, as Franklin D. Roosevelt said, 'the Worst' of the Intellectuals?,' paper read at joint session of Society for Philosophy and Technology and Society for the Advancement of American Philosophy, Washington, DC, 27 December 1985.
[2] Feuer, *op. cit.*, p. 14.
[3] *Ibid.*, pp. 38, 43.
[4] 'Philosophies of Freedom,' originally published in *Freedom in the Modern World,* ed. Horace Kallen (New York, 1928), reprinted in *On Experience, Nature and Freedom,* ed. Richard J. Bernstein (Indianapolis/New York: Bobbs-Merrill, 1960), pp. 271–272.
[5] 'Professional Spirit among Teachers,' *American Teacher,* October 1913, pp. 114–116; 'Professional Organization of Teachers,' *American Teacher,* September 1916, pp. 99–101; 'Why I Am a Member of the Teachers Union,' in *American Teacher,* January 1928, p. 4.
[6] George Dykhuizen, *The Life and Mind of John Dewey,* ed. Jo Ann Boydston with introduction by Harold Taylor (London and Amsterdam: Feffer & Simons; Carbondale and Edwardsville: Southern Illinois University Press, 1973), pp. 256–258.
[7] Dewey, 'The Manufacturers' Association and the Public School,' *Journal of the National Education Association* 17 (1928): 61–62. See Hewlett (note 10, below),

p. 119; Dykhuizen, p. 231.
[8] Dykhuizen, *op. cit.,* pp. 231–232, 239; Hewlett (note 10, below), p. 119.
[9] Dewey, 'Labor Politics and Labor Education,' *New Republic,* 2 January 1929, p. 213.
[10] Charles F. Hewlett, *Troubled Philosopher: John Dewey and the Struggle for World Peace* (Port Washington, NY, and London: Kennikat Press, 1977), pp. 119, 121–122; Dykhuizen, *op. cit.,* p. 230.
[11] Dykhuizen, *op. cit.,* p. 229. These same concerns Dewey also publicized in the pages of the *New York Times:* 'Asks Hoover to Act on Unemployment' (21 July 1930, p. 17); 'Asks Federal Fund to Aid Unemployed' (12 May 1930, p. 35); 'Puts Need of Idle at Two Billions' (26 October 1930, p. 21).
[12] Dykhuizen, *op. cit.,* pp. 167–169, 304–307; Hewlett, *op. cit.,* p. 33. See Dewey, 'Professional Freedom,' *New York Times,* 22 October 1915, p. 10; 'The Case of the Professor and the Public Interest,' *Dial* 62 (1917): 435–437. Dewey was also involved in the defense of secondary school teachers.
[13] *Democracy and Education* (New York: Macmillan, 1916), preface, p. 5. (Cited hereafter as DAE.)
[14] *Liberalism and Social Action* (New York: Capricorn, 1963; original, 1935), p. 52. (Cited hereafter as LSA.)
[15] *Ibid.,* p. 47.
[16] *Logic: The Theory of Inquiry* (New York: Holt, Rinehart & Winston, 1938), p. 163. (Cited hereafter as LTI.)
[17] LSA, pp. 47, 71.
[18] *Essays in Experimental Logic* (Chicago: University of Chicago Press, 1916; reprinted New York: Dover, 1954), p. 414.
[19] *Ibid.,* p. 306.
[20] LTI, p. 492.
[21] LSA, pp. 72, 74, 75, 82, 83–87.
[22] *Ibid.,* pp. 92, 90.
[23] *Ibid.,* p. 38; see also pp. 35–37.
[24] *Ibid.,* p. 90; see also LTI, pp. 504–505.
[25] LSA, p. 54.
[26] DAE, p. 98.
[27] LSA, p. 38. Years before, at the time of World War I, Dewey had directed similar criticisms against the elitism of both the German and the British educational systems. See Gary Bullert, *The Politics of John Dewey* (Buffalo, NY: Prometheus, 1983), pp. 23–24.
[28] *Ibid.,* pp. 20–21.
[29] *Ibid.,* p. 25.
[30] *Ibid.,* pp. 61, 64.
[31] Arthur Lothstein, 'Salving from the Dross: John Dewey's Anarcho-Communalism,' *The Philosophical Forum* (Boston), 20 (Fall 1978): 98.
[32] *Ibid.,* p. 15–16.
[33] *Ibid.,* p. 73. Dewey himself considered *Democracy and Education* (1916) to be pivotal to an understanding of his views on the relation between science and morality, but regretted that philosophers, unlike teachers, seldom read it. See *On Experience, Nature and Freedom,* ed. Richard J. Bernstein (Indianapolis/New York: Bobbs-

Merrill, 1960), pp. 14–15.

[34] *The Quest for Certainty* (New York: Capricorn, 1960; original, 1929), pp. 198–200. See also *Experience and Nature* (2d ed.; LaSalle, IL: Open Court, 1929; original, 1925), pp. 128–129. Special thanks to Larry Hickman and an anonymous reviewer for goading me to clarify this point.

[35] LTI, pp. 104–105ff.

[36] *Ibid.*, pp. 75–76.

[37] LSA, p. 31.

[38] LTI, p. 73. Thus did he encourage teachers to organize in order to bring 'brainworkers' together with 'handworkers' in 'service to the general public': 'Professional Organization of Teachers' (note 5, above).

[39] Don Colburn, 'Physician, Organize Thyself,' *The Washington Post National Weekly Edition*, 12 August 1985.

[40] Matthew W. Finkin, Robert A. Goldstein, and Woodley B. Osborne, *A Primer on Collective Bargaining for College & University Faculty* (Washington, DC: American Association of University Professors, 1975), p. i.

[41] Scott Heller, 'Faculty Unions Still Growing, Study Finds,' *The Chronicle of Higher Education*, 8 May 1985, pp. 1, 24; National Center for the Study of Collective Bargaining in Higher Education and the Professions, Baruch College, New York, NY, 1985.

[42] Everett Carll Ladd, Jr., and Seymour Martin Lipset, *Professors, Unions, and American Higher Education* (Berkeley, CA: The Carnegie Commission on Higher Education, 1973), p. 11.

[43] *Ibid.*, p. 103.

[44] Robert L. Jacobson, 'Low Pay and Declining Working Conditions Seen Threatening Colleges' Teacher Supply,' *The Chronicle of Higher Education*, March 27, 1985, p. 21; *On Campus*, December 1985/January 1986, p. 3. See Michael I. Sovern, *Passing the Torch: Graduate Education and America's Future*, Columbia University President's Report, 1985.

LARRY HICKMAN

DOING AND MAKING IN A DEMOCRACY: DEWEY'S EXPERIENCE OF TECHNOLOGY

> Understanding has to be in terms of how things work and how to do things. Understanding, by its very nature is related to action; just as information, by its very nature, is isolated from action ... only ... by accident.
> JOHN DEWEY[1]

Advancing a claim that was then regarded as radical, and is still widely misunderstood, John Dewey argued that most of his philosophical predecessors, even those who had claimed the methods of science as their own, had been guilty of a failure to recognize the importance of technology. He suggested that this was due in part to their prejudice against the impermanent materials utilized by artisans and craftspeople, in part to their tendency to deprecate the social classes whose members have traditionally dealt with doing and making in the practical sphere, and in part to their rejection of what he took to be the democratizing tendencies of technological methods.

Analysis of the web of technical artifacts and methods which mankind weaves and in which it lives and works was for Dewey a lifelong task.[2] His early work, between 1892 and 1898, exhibits a preoccupation with the relations between the sciences and the industrial arts, and between what were then known as 'normal' and 'technical' schools. His middle work, from 1899 to 1924, contains discussions of the ways in which intelligence is related to the use of technological artifacts, and of the ways in which concrete tools such as agricultural implements are related to tools that are less tangible, such as logical connectors.

Dewey's later work, from 1925 until his death in 1952, including *Experience and Nature* (1925) and *Art as Experience* (1934) developed these themes in detail. He articulated an account of the philosophical implications of technology during its classical, modern, and contemporary periods, and he anticipated many of the issues and debates which now occupy those working in the emerging field of the philosophy of technology. In this connection, chapters four and five of *Experience and Nature,* central chapters in what is regarded by many as Dewey's most important work, are devoted almost entirely to a critique of technology. Moreover, those of his later works that focus on science, education, religion, and democracy are richly furnished with examples and metaphors from the technical sphere.

It was Dewey's contention that his philosophical predecessors had for the most part mislocated technology with respect to science, metaphysics, and social thought. Plato and Aristotle, each in his own distinctive way, had attempted to relocate technology outside the work of the artisan and outside the sphere of human interaction with changeable matter. Plato, especially in the *Timaeus,* did this by establishing a kind of 'grand artisan' outside the realm of nature. Aristotle did so by making nature itself the grand artisan whose task it is to establish fixed ends fit to be contemplated as ends-in-themselves, not as instruments for further ends.

What resulted was not just a perversion of technology itself, but a stunting of the growth of science and social inquiry as well. *The Republic* richly documents the consequences for social thought in general and for democracy in particular of this turn against experience in its full-bodied sense. It is there that Plato relegates *techne,* the activities of the technical artisan, to the lowest rung of his socio-political hierarchy, and at the same time characterizes an attenuated and immaterial form of *techne,* that of the totalitarian social engineer, as the purest and most important of social activities.

It was Dewey's contention that Plato had placed the artisan at the bottom of the social hierarchy for the same reason that he had so adamantly demanded censorship of the work of the plastic and dramatic artists: the methods of *techne* are too powerful to be left in the hands of artists and craftsmen. Unhindered by the repressive legislation of the perfect guardians, the practioners of *techne* in its concrete sense would have proved a threat to the 'thinkers' of the Republic.

As for Aristotle, it was Dewey's view that the *Politics* fosters a view of the city state so constructed that its justification rests on ends 'given' by nature. The activities of the practitioner of *techne* are, as they were in Plato, refined and sublimated. For Aristotle, however, the beneficiary of his transference is not a system of supernature contemplated by the philosopher king but nature itself, which becomes the grand artisan. As Plato had, but less perniciously so, Aristotle plundered much of the work of the artisan of its creative and social significance and relocated its content elsewhere.[3]

Dewey read the Greek attitudes toward science as part and parcel of their unfortunate attitude toward *techne.* He argued that the Greeks' abhorrence of the mutability inherent in the tasks and materials of technology had led to a science of 'demonstration,' to a science of contemplation, to an attempt to possess something already finished, 'out there' and complete. In fact, they had invented not so much science as the idea of one. He issued the warning that when inquiry is focused in the sphere of objects esteemed for their own intrinsic qualities, whether that sphere be cast as supernatural or extranatural, as it was for Plato, or natural and immanent but complete, as it was for Aristotle, then whether that inquiry concerns itself with materials and artifacts, conceptual models of nature, or the ways in which social organization takes place, such inquiry will fail to increase our knowledge of things as they are.

As for modern science, the science of Copernicus, Galileo, Kepler, and Newton, it was Dewey's view that its advances were attributable more to what its practitioners were *doing* than to what they *thought* they were doing. It was not that what was novel in its theories did not advance its practice; it was simply that its metatheory, its metaphysics and epistemology, more often than not failed to grasp what was innovative in those first order theories. Dewey had high praise for the new mathematical techniques of substitution, and suggested that they constituted a 'system of exchange and mutual conversion carried to its limit.'[4] The objects of science thus became 'amenable to transformation in virtue of reciprocal substitutions.'[5] But the metaphysics and epistemology of the new science were still wedded to the old ideas of a finished universe.

Dewey's claims in this regard are by no means uncontroversial. Desmond Lee, for example, in the introduction to his translation of the *Timaeus,* rejects the view that the contempt held by the Greeks for the work of the artisan discouraged experimentation and hindered the development of technology. He argues that the aristocracy of seventeenth century England, a time and place of enormous technological development, had at least as much disdain for the artisan as had the Greeks. He further suggests that inhibitions of technological development in the classical world were not always aristocratic. The Roman contractor, whom he identifies as a 'fairly rough type, often a freedman,'[6] would certainly have been glad to have profited from technological development if it had been possible for him to do so. Instead, Lee suggests, the weakness of ancient metallurgy and a lack of precise instrumentation were among the inhibiting factors. But why should these technological materials and instruments not have developed? Lee suggests that there was a conceptual reason: the Greeks had tied science to philosophy, and 'philosophy is concerned to understand rather than to change.'[7] For Lee, the contribution of Galileo to the advancement of experimental science was that he took the technical tools and artifacts available to him, tools and artifacts that had gradually become much more sophisticated than those developed by the Greeks, and used them to 'untie' science from philosophy.

Dewey repeatedly rejected any view of philosophy that had as its goal understanding without change, for he thought that understanding of any legitimate sort entails change. He also argued that hope of financial gain, even by the most 'rough and ready' of contractors, is in itself insufficient to promote technological development. It may in fact thwart or divert such development. On one point at least, his position is consonant with that of Lee. They agree that experimental science requires active transaction with environing conditions. But for Dewey, those among his predecessors and those among his contemporaries who could not see that this is also true of philosophy had not fully appropriated the lessons of the scientific revolution of the seventeenth century.

Put in the socio-political terms by means of which he had analyzed the fate of *techne* among the Greeks of the fourth century B.C.E., it was Dewey's contention that the new science of the seventeenth century exhibited a surge of democratic methods and assertion of adaptive practice. But there was a broad gulf between what the new science said it was doing and what it was actually about.

The official view of what modern science was about was still conservative and authoritarian. Its apologists continued to traffic in antecedent truths, demonstrations, and certitudes. They held fast to what we today call foundationalism and the correspondence theory of truth. Moreover, its metatheory continued in this vein long after the new science had enjoyed the prodigious successes that resulted from its practical commitment to the treatment of natural ends as instruments for further inquiry and transaction with nature, rather than as fixed objects of contemplation. Thus did the practice and first order theories of seventeenth century science relocate technology *de facto* in terms of its new spirit of practical experimentalism, even if its metatheory did not do so. Dewey thought that the genius of the new science was its discovery that 'knowledge is an affair of *making* sure, not of grasping antecedently given sureties.'[8]

Nevertheless, many of the metatheorists of seventeenth century science, among them the most respected metaphysicians of the time, demonstrated an ignorance that its taproot was in practice, that is, in a transaction with the growing body of tools and artifacts that made the new science possible.

Contemporary historians of science and technology continue to commit this error. The following description by Daniel Boorstin of the work of Galileo the telescope-maker is illustrative of this mistake. 'With no special insight into the science of optics,' Boorstin writes:

> Galileo, a deft instrument-maker, had made his device by trial and error. But if Galileo had been merely a practical man, the telescope would not have been such a troublemaker.[9]

Dewey would have found in this characterization a vestige of the very mistake made by most of the early philosophers of modern science. His view was that Galileo was not so much proceeding by means of trial and error as he was 'thinking' with his materials, inquiring into their possibilities in a way that had much more in common with the activities of the artists and craftsmen of Plato's Greece than the philosophers of the modern period realized. Dewey's view of what Galileo was doing is closer to the description of his activities provided us by Paolo Rossi:

> Kepler was to lay the foundation of the new optics in the *Paralipomena* of 1604, but it was to be a scientist-technician like Galileo who was to muster up the courage 'to look' by using the telescope. He skillfully transformed a use-object which had progressed only 'through practice,' partly accepted in military circles but ignored by the official scientific establishment, into a powerful instrument of scientific exploration.[10]

Dewey argued that inquiry into materials such as that practiced by Galileo precedes and conditions inquiry of a more conceptual variety. It also informs its methodology, and terminates its activity in further concrete application. This was perhaps Dewey's most important contribution to the debates concerning the relations between science and technology. Even though the craftsman who thinks in and with materials may not translate that thought into the conceptual sphere, and conversely even though those who think by means of conceptual tools are frequently unable to bring their work to fruition in practical terms, there is nevertheless no reason to posit a methodological gap between the two enterprises.

It is only the infelicitous social prejudice regarding the media in which inquiry is undertaken, the misunderstanding of the talents and dispositions of those who direct the inquiry, and the unfortunate social and cultural boundaries assumed to exist between those modes of inquiry that perpetuate the appearance of a gap that is not in fact justified from the standpoint of methodology. In short, intelligence with respect to materials is fully the equal of intelligence with respect to those enterprises we normally consider 'conceptual': for example, science and social thought. Not only are their methods basically the same, but it is only by the cooperation of each with the other that human knowledge is advanced.

Dewey made this point forcefully in *Art as Experience*. 'Any idea,' he wrote,

> that ignores the necessary role of intelligence in production of works of art is based upon identification of thinking with use of one special kind of material, verbal signs and words. To think effectively in terms of relations of qualities is as severe a demand upon thought as to think in terms of symbols, verbal and mathematical. Indeed, since words are easily manipulated in mechanical ways, the production of a work of genuine art probably demands more intelligence than does most of the so-called thinking that goes on among those who pride themselves on being 'intellectuals.'[11]

He was careful to include the 'practical' or 'technological' arts in this characterization. 'Art,' he suggested, 'denotes a process of doing or making. This is as true of fine as of technological art.'[12]

It was Dewey's claim, then, that philosophy during its modern period, from the seventeenth to the nineteenth centuries, failed to locate technology properly because its allegiance was still tied to the metaphysics of contemplation, of antecedent truths, demonstration, and certitude. But his analysis did not simply take the part of the Empiricists against the Rationalists. Some modern philosophers, he pointedly reminded us, surrendered the antecedent truths of reason only to accept antecedent truths of sensation. Modern Empiricism, according to his view, committed itself to an equally egregious form of foundationalism.

For the bulk of philosophy in its modern period, nature was thought to be a vast machine. Living in the shadow of Darwin as he did, Dewey rejected the machine as leading metaphor and replaced it with the organism. But even to those who have transcended the metaphor of world-as-machine there is still the fact of machines, and the problem of how to relate to them. A machine can be contemplated as something finished, and its workings discovered and admired. Further, it can be examined as something complete but in need of occasional repair. Or it can be interacted with as something ongoing, unstable and provisional, as a tool which is utilized for enlarging transactions of self and society with environing conditions. It was Dewey's contention that the discussions of the nature of the world-as-machine in the seventeenth and eighteenth centuries were primarily focused on the first two of these attitudes. Of course each of these three possibilities involves some level of interaction with nature. But it is only with the third that there comes to be genuine transaction with nature, awareness of such transaction, and inclusion of that awareness in the metatheories of science.

In the political sphere, of course, it was a great advance over the old supernaturalist and extranaturalist views to think of the world as repairable, even if it was not yet fully open to transaction. In *Liberalism and Social Action,* Dewey praised the advances made by Bentham on just these grounds. But he also warned of treating the world-machine as *merely* examinable and repairable. He cautioned against Bentham's acceptance of mankind as 'a reckoning machine.'[13] The old machine metaphors of Bentham and others neglected the fact that relations between human beings and their political environments are always 'relations of ongoing affairs characterized by beginnings and endings which mark them off into unstable individuals.'[14] These individual relations are in need of continual and intelligent reevaluation and reconfiguration by means of practical inquiry.

Failure to make this conceptual shift from machine as finished though imperfect and repairable to machine as incomplete and unstable instrument has precipitated in our time a situation well described by Stuart Hampshire in a polemic against Utilitarianism, a cluster of positions against which Dewey also argued. In its emphasis on repair as opposed to transaction, Hampshire suggested, much recent thought has led to

> new abstract cruelty in politics, a dull, destructive political righteousness: mechanical, quantitative thinking, leaden academic minds setting out their moral calculation in leaden abstract prose, and more civilized and more superstitious people destroyed because of enlightened calculations that have proved wrong.[15]

It might be objected that it was during this modern period of science that the United States of America, the most influential democracy of the contemporary world, was founded. It might further be argued that among the framers of the political documents of that infant democracy were deists, practitioners of a form of religious faith that explicitly regards God as artisan and a virtually finished universe as His handiwork. But among these social experimenters were gadget-makers, mechanics, and tinkerers. Thomas Jefferson, whom Dewey greatly and publicly admired, consistently spoke of political and social experimentation in a manner that echoed his transaction with clocks, agricultural methods, and gadgets of many diverse sorts. Jefferson repeatedly referred to the government which he helped establish as an *experiment,* moreover one whose institutions and laws would be in need of recurring modification by each succeeding generation. For government, as for nature, contemplation had been replaced by examination, and that in turn by experimentation whose goal was constant attention to possibilities of adjustment and amelioration.

Jefferson's transactionist orientation to technology, to social thought, and to the broader world of his experience stands in stark contrast to the examinationist program of Descartes a little over a century earlier. L.J. Beck gives the following account of Descartes' attitude toward Galileo's telescope as exhibited in his *Dioptrique* of 1637 and his correspondence with Ferrier:

> Already at La Flèche, Descartes had probably heard of the discoveries made by Galileo through the use of the telescope. Descartes wishes to draw up a plan for the construction of an even better one, and above all of a more powerful lens. This cannot be done until, he tells us, it is known, what happens when light traverses several lenses, until the law of refraction has been established and the problem of the *linea anaclastica* solved. Then only can the plan of the various curves of the lenses be worked out. Descartes works these out but, as one can see, the unfortunate Ferrier is unable to carry out in practice the difficult requirements set by Descartes. Galileo cannot solve the problem of the *linea anaclastica;* he does not know the law of refraction, but he manages to construct an instrument which gives a substantial magnification. Kepler, slightly more theoretical, knew only of approximations to the law of refraction but his description of the telescope provided a working model for future astronomers. Descartes required the exact measurements for his lenses, and failing this, he lost interest in the whole topic.[16]

It was precisely this debate between the transactionist craftsmen-practitioners of modern science and technology and those seeking to examine its hypothetical and metaphysical foundations that was a matter of intense interest to Dewey. It provided evidence for his thesis that metaphysicians of the period had mislocated the place of technological practice. Writing of the controversy between the Cartesian school and that of Galileo and Newton, he lauded the triumph of the latter because of its emphasis on 'experience.' And his characterization of 'experience' made extensive use of examples of inquiry in the technological sphere.[17] In a rather sad aside he suggested that, 'We may, if sufficiently hopeful, anticipate a similar outcome in philosophy. But the date does not appear to be close at hand; we are nearer in philosophic theory to the time of Roger Bacon than to that of Newton.'[18] Dewey wanted to locate technology in a realm that is neither supernatural nor extranatural, an organic realm in which the only telic elements are those of the natural ends of objects, individuals, and events, all of which in turn may be utilized as means to further ends. It was his view that the legitimate place of technology is alongside science and social thought as one of several branches of inquiry. On his reading, technology is not inferior to its brother and sister branches, and may in some respects even be more important than they in that its unique qualities serve to inform, enhance, and promote those siblings in ways that they are incapable of reciprocating.

What are these unique qualities? I have already alluded to his commitments to what Don Ihde would later call 'the historical-ontological priority of technology over science.'[19] In 1925 Dewey argued the historical component of this claim when he suggested that in spite of the obvious fact that

> the sciences were born of the arts – the physical sciences of the crafts and technologies of healing, navigation, war and the working of wood, metals, leather, flax and wool; the mental sciences of the arts of political management, ... it is still commonly [and erroneously] argued that technology is merely 'applied science.'[20]

He further argued that modern science

> represents a generalized recognition and adoption of the point of view of the useful arts, for it proceeds by employment of a similar operative technique of manipulation and reduction. Physical science would be impossible without the appliances and procedures of separation and combination of the industrial arts.[21]

In addressing the 'ontological' component of his claim, Dewey reminded us that what is peculiar to human interaction with the world is not its enjoyment, but the necessity of grappling with it at the technological level and the knowledge, or science, which follows upon that interaction. In Dewey's words, 'It was not enjoyment of the apple but the enforced penalty of labor that made man as the gods, *knowing* good and evil instead of just having and enjoying them.'[22] To contrast knowing and having, as Dewey did in this remark, is to allude to his treatment of knowledge as hypothesis, pointing to an unfinished future in which both inquiring human beings and their environments undergo alterations.

This is a view that would reemerge in Heidegger's essay, 'The Question Concerning Technology.' Technology is there differentiated into (1) technology as a tool of science, (2) technology as the activities of the craftsman *(techne)* and (3) technology in its ultimate sense, *aletheia* or revealing. It is this last sense of technology that is most basic to Heidegger's account: 'Instrumentality is considered to be the fundamental characteristic of technology. If we inquire, step by step, into what technology, represented as means, actually is, then we shall arrive at revealing. The possibility of all productive manufacturing lies in revealing.'[23] Again, 'Technology is a mode of revealing. Technology comes to presence ... in the realm where revealing and unconcealment take place, where aletheia, truth, happens.'[24]

I have recalled Heidegger's account of the ontological priority of technology over science because of the light it sheds on Dewey's. For Dewey, it is technological instrumentality (what Heidegger calls 'revealing') that characterizes the most primitive relation between the activities of men and women and the world of their experience. Such instrumentality ties together the myths that tell of the manner in which labor entered the world and the myths that constitute our most up-to-date theories of political economy.

But if technology is prior to science both historically and ontologically, it also is responsible for the prestige enjoyed by science. Dewey argued that the successes of science have been due not so much to what he called 'scientific temper' as to 'scientific technique.' In his essay 'Human Nature and Scholarship,' Dewey argued that,

> Scientific technique, as distinguished from the scientific temper, is concerned with the methods by which matter is manipulated. It is the source of special technologies, as in the application of electricity to daily life; it is concerned with immediate fruits of a practical kind in a sense in which *practical* has a special and technical meaning – power stations, broadcasting, lighting, the telephone, the ignition system of automobiles.[25]

Further,

> The inherent idealism of the scientific temper is submerged, for the mass of human beings, in the use and enjoyment of the material power and material comforts that have resulted from its technical applications.[26]

What was Dewey's view of the location of technology with respect to epistemology? 'Knowledge ceases to be a mental mirror of the universe and becomes a practical tool in the manipulation of matter.'[27] Dewey reiterated his radical position in *Experience and Nature:* 'In the practice of science, knowledge is an affair of *making* sure, not of grasping antecedently given sureties.'[28]

Dewey not only viewed technology as the primary means of inquiry open to those individuals cut off from what normally goes on in laboratories, observatories and places of special research; he suggested that technology was a special avenue of inquiry open to those individuals living in closed societies where social inquiry is suppressed. But he was neither idealist nor utopian. He knew that even in open societies there would be those who prefer appeals to tenacity, authority, or the *a priori* to free and open inquiry as methods of fixing their beliefs. It was with this in mind that he suggested that technology may also operate as a buffer between the forces of anti-science and those of science.

I do not think that he would have been surprised that those who now attempt to promote the teaching of a literal reading of the Genesis myth of creation do so while claiming to march under the banner of science. The advances of science propagated in the technological sphere have mooted many of the old anti-scientific arguments, or at least required that they be masked in scientific jargon. Dewey repeatedly demonstrated his conviction that the work of those who take the pluralistic values of free and open inquiry seriously will never be finished. This is a central aspect of his philosophy of education.

But if technology for Dewey forms a buffer between the forces of anti-science and those of science, it also functions as a means by which science may be appropriated by the scientifically uninformed. There are two ways in which this takes place: not only have the 'fruits' of technology become ubiquitous, but the methods of science, historically and ontologically dependent on technology, may be reintroduced into the field of technological practice and use with new authority. But with these two outcomes of technology, one immediate the other mediate, come two dangers. The first is the one indicated by Jose Ortega y Gasset, that technological men and women may become like the aboriginal forest or jungle dweller, just 'picking' the technological fruits as if they had been supplied by a 'natural' system beyond their understanding or control. The second is that technology will once again become mislocated with respect to science, that is, that it will once again suffer the deprecation it underwent during the period of classical, modern, and much of contemporary philosophy. These are dangers to social organization in general, and to democracy in particular, because they signal the truncation of the full spectrum of inquiry necessary to the transaction of human beings with their environment and the consequent knowing of things as they are and can be.

In his 1944 essay, 'Democratic Faith in Education,' Dewey made even more explicit his concern regarding the dangers to technology and to social inquiry. He could have been writing of our current situation when he characterized the first of these dangers as that of 'laissez faire naturalism.' To those who would appeal to the forces of the 'invisible hand' or 'the [undefined] laws of the marketplace,' Dewey had this to say:

Technically speaking the policy known as *laissez-faire* is one of limited application. But its limited and technical significance is one instance of a manifestation of widespread trust in the ability of impersonal forces, popularly called Nature, to do a work that has to be done by human insight, foresight and purposeful planning.[29]

He applied this assessment to planning in national and international affairs, suggesting that the refusal to apply the methods of the technological sciences in those areas had led to a state of imbalance and 'profoundly disturbed equilibrium.'[30]

A second danger to technological and social inquiry lies in the attitudes and activities of the 'humanist' who attacks technology as 'inherently materialistic and as usurping the place properly held by abstract moral precepts.'[31] He suggested that these moral precepts had remained abstract precisely because those defining them had divorced ends from the means by which they are to be realized.

Dewey specifically criticized the Hutchins program, which set out to separate technical training from liberal education, a situation which has both become a fact of our universities in the 1980s and is seen by many as tending to create a permanent social wound as it becomes more widely practiced in our secondary schools. When the putrefaction of the wound is eventually discovered, the blame will likely be located at the door of 'technology,' and not in its proper place, namely, the failure to apply the inquiry which is characteristic of experimental science and technology at its best to all areas of human endeavor.

In that same essay Dewey mentioned a third threat to technological and social inquiry. It is consequent on the activities of contemporary Luddites, especially of the theocratic variety. They resist the application of scientific and technical methods to the field of human concerns and human affairs both because they tend to think of themselves as outside of and above nature, and because they prefer a return to the medieval prescientific doctrine of a supernatural foundation and outlook in all social and moral matters. He further suggested that this group erroneously believes that the methods of science and technology have been applied to every area of human concern – and found wanting.

A special variety of this third faction has become even more militant than it was at the time Dewey issued his warning. Both anti-scientific and anti-democratic, Christian fundamentalists of the extreme right have nevertheless adopted the tools of electronic technology to advance their aims. Dewey was aware that technological advances could be appropriated by authoritarian forces, and discussed this phenomenon in detail in his 1936 essay, 'Religion, Science and Philosophy.'

The two forces that particularly concerned him in that dangerous year, a year which saw the growing power of fascism throughout the world, were what he called 'political nationalism' and 'finance-capital,' both important allies of the religious right in our own time. He called these movements 'new religions.' They, like religious fundamentalism, depend on the *a priori* and the revealed as a substitute for intelligent inquiry. They, like religious fundamentalism, have their established dogmatic creeds, their fixed rites and ceremonies, their central institutional authority, their distinction between the faithful and the unbelievers, with persecution of heretics who do not accept the true faith.[32]

They share a further important characteristic of religions, *viz.*, they dote on the *terminus a quo* rather than on the *terminus ad quem*, a doctrine of 'original intent' rather than a careful attention to consequences. *Experience and Nature* warns of the capture of applied science by these elements: they work to channel it toward 'private and economic class purposes and privileges. When inquiry is narrowed by such motivation or interest, the consequence is in so far disastrous both to science and to human life.'[33] Dewey reminded us that these potential disasters are not due to the practical nature of technology, but to the defects and perversions of morality as it is embodied in institutions and the effects of such institutions upon personal disposition.

This essay would remain incomplete were I not to quote an extended passage from an address delivered by Dewey at a celebration of his eightieth birthday. There is perhaps no more eloquent characterization of the interaction between inquiry at its various technological, scientific, and political levels in all of Dewey's writings. 'Democracy' is there characterized as

a belief in the ability of human experience to generate the aims and methods by which further experience will grow in ordered richness. Every other form of moral and social faith rests on the idea that experience must be subjected at some point or other to some form of external control; to some 'authority' alleged to exist outside the process of experience. Democracy is the faith that the process of experience is more important than any special result attained, so that special results achieved are of ultimate value only as they are used to enrich and order the ongoing process. Since the process of experience is capable of being educative, faith in democracy is all one with faith in experience and education. All ends and values that are cut off from the ongoing process become arrests, fixations. They strive to fixate what has been gained instead of using it to open the road and point the way to new and better experience.

If one asks what is meant by experience in this connection, my reply is that it is that free interaction of individual human beings with surrounding conditions, especially the human surroundings, which develops and satisfies need and desire by increasing knowledge of things as they are. Knowledge of conditions as they are is the only solid ground for communication and sharing; all other communication means the subjection of some persons to the personal opinion of other persons. Need and desire – out of which grow purpose and direction of energy – go beyond what exists, and hence beyond knowledge, beyond science. They continually open the way into the unexplored and unattained future.[34]

Texas A&M University

NOTES

[1] John Dewey, 'The Challenge of Democracy to Education,' in *Problems of Men* (PM) (New York: Philosophical Library, 1946), p. 49.

[2] Dewey devoted no single work to his analysis of technology. His treatment of the subject is diffused throughout the more than 15,000 pages of his published work. For this and other reasons, the secondary literature of this field is scant. There are, however, a few brave essays on the subject, including an important and interesting one by Webster F. Hood, 'Dewey and Technology,' in P. Durbin, ed., *Research in Philosophy and Technology,* vol. 5 (Greenwich, Conn.: JAI Press, 1982), pp. 189–207.

[3] As Hannah Arendt pointed out in her classic essay 'The *Vita Activa* in the Modern Age,' in *The Human Condition* (Chicago: University of Chicago Press, 1958), Aristotle's *Metaphysics* had placed the sciences of fabrication above the practical sciences but below the theoretical ones. She thought that this was so because for the Greeks the contemplation of the theoretical sciences was thought to be an inherent element in fabrication, that is, what allowed the craftsman to judge the finished product. The point of Dewey's critique of the Greeks was not that they did not give *techne* some of its due, but that they thought it secondary to grander technical forces, viz., supernature for Plato, and nature for Aristotle.

[4] John Dewey, *The Later Works* (LW), ed. Jo Ann Boydston (Carbondale and Edwardsville: Southern Illinois University Press, 1981–), 1:115.

[5] *Ibid.*

[6] Plato, *Timaeus and Critias,* translated and with an introduction and an appendix by Desmond Lee (New York: Penguin Books, 1983), p. 18.

[7] *Ibid.*, p. 19.

[8] Dewey, *The Middle Works* (MW), ed. J. Boydston (Carbondale and Edwardsville: Southern Illinois University Press, 1976–1983), 1:123. Dewey's emphasis.

[9] Daniel Boorstin, *The Discoverers* (New York: Random House, 1983), p. 318.

[10] Paolo Rossi, *Philosophy, Technology and the Arts in the Early Modern Era* (New York: Harper and Row, 1970), p. 35.

[11] LW 10:52.

[12] LW 10:53.

[13] John Dewey, *Liberalism and Social Action* (New York: Putnam's, 1980; original, 1935), p. 30.

[14] LW 1:127.

[15] Stuart Hampshire, 'Morality and Pessimism,' in R. Beehler and A. Drengson, eds., *The Philosophy of Society* (London: Methuen, 1978), p. 32.

[16] L.J. Beck, *The Method of Descartes* (Oxford: Clarendon Press, 1952), pp. 261–262.

[17] LW 1:14–15.

[18] LW 10:15.

[19] Don Ihde, *Existential Technics* (Albany: State University of New York Press, 1983), pp. 25–46.

[20] LW 1:105.

[21] LW 1:108.

22 LW 1:100.
23 Martin Heidegger, *The Question Concerning Technology and Other Essays,* trans. and ed. William Lovitt (New York: Harper and Row, 1977), p. 12.
24 *Ibid.,* p. 13.
25 John Dewey, 'Religion, Science and Philosophy' (RSP), in *Problems of Men* (PM) (see note 1, above), p. 173.
26 *Ibid.,* p. 174.
27 *Ibid.,* p. 172.
28 LW 1:123.
29 John Dewey, 'Democratic Faith in Education,' in PM, p. 25.
30 *Ibid.,* p. 27.
31 *Ibid.*
32 RSP (see note 25, above), p. 175.
33 LW 1:130–131.
34 John Dewey, 'Creative Democracy: The Task Before Us,' in *John Dewey,* ed. Irwin Edman (Indianapolis/New York: Bobbs-Merrill, 1955), p. 314.

JOSEPH MARGOLIS

PRAGMATISM, *PRAXIS*, AND THE TECHNOLOGICAL

It is difficult to explain the sense in which a theory of technology constitutes a theory of knowledge or inquiry – the sense in which it is not merely a theory of contingent interests or objectives that *an already cognitively competent society* applies its powers to. In a large sense, 'technology' signifies the conditions of existence (the 'existential' conditions) under which human beings minimally function as cognitively competent. In this regard, a theory of technology collects: (1) the empirically specifiable skills humans exhibit in manipulating and altering actual things relative to the tacit conditions of survival as well as to the explicit testing of particular claims; (2) the nesting of these first-order skills within second-order reflections about the matching of cognitive powers and cognized world congruent with man's empirical achievements; and (3) the nesting of that nesting in further reflections on what kind of existent being man (generic sense) must be in order to accomplish what we take to be accomplished in (1) and (2). A theory of technology seeks to penetrate to an understanding of what in the very nature of man: (a) makes it possible for him to achieve a science *vis-à-vis* the things of the world, and (b) also accounts for *that* capacity's being effective within whatever preformative conditions we suppose it to function. The theory of technology is the theory of how man's distinctive mode of existence both enables and constrains his effective science. To address the question need not be to pretend to any privileged or hierarchically ordered forms of cognitive access to information about any of the issues (1–3).

To entertain the speculation, under the condition that nature is not cognitively transparent, that man has no cognitively privileged entry with respect to fixing the structures of the world thus cognized, and also under the condition that that achievement *is* the natural flourishing of the way in which the human species survives in the world, is to address the question as a *pragmatist*. Also, stretching terminology somewhat beyond its usual usage: to address the question as a pragmatist, under the further condition that we are searching for a conception of how, primordially, man engages the world in a cognitively effective way (how his science is preconditioned or preformed) is also to address the question as a *phenomenologist* (in a deliberately neutral sense neatly shared by such otherwise disparate figures as Peirce and Heidegger).[1] The theory of technology may be fairly construed as an exercise of this jointly phenomenological and pragmatist sort – without disallowing any otherwise eligible sense of the notion. But to appreciate the distinctive power of its special contribution, one must construct a kind of conceptual narrative (doubtless, alternative such narratives) by the dialectical use of which one may disclose the increasing aptness of a series of hierarchically placed but reciprocally affected inquiries. On review, it appears that *the same powers of inquiry* are in play in moving from empirical science to the theory of knowledge to the phenomenology of man – and that those powers cannot fail to be manifestations of the 'technic' or technological way of existing.

Our purpose is to show the need for a certain kind of inquiry about the human condition that, though conceptually identified as an essential precondition of our cognitive achievements, is nevertheless so identified only by reflexive reference to whatever we take to be the thus encumbered achievements themselves. Every putative encumbrance, therefore, for example the limits of an historical horizon or the determining function of different forms of economic production or the like, will, inevitably, exhibit a dual role: it will color *both* the development of a philosophical anthropology fitted to the entire range of human sciences and to the bearing of those sciences upon the natural sciences *and* the formation of a phenomenological or ontological conception of what in the very nature of human existence could serve as the ground or precondition for those effective sciences and that effective philosophical anthropology. Within Western philosophy and in a pertinently pragmatist spirit, Peirce's theory of signs or Habermas's theory of communicative competence may perhaps be taken as strong specimens of what a philosophical anthropology may look like; and Heidegger's theory of technology as a pervasive form of man's existential power to disclose and to have disclosed to him a world suitably structured for cognitive inquiry or Gadamer's theory of the fusion of horizons along related lines may perhaps be taken as strong specimens of what a phenomenology may look like. One suspects, therefore, that such distinctions are no more than heuristically favored or hierarchically ordered at particular moments of systematic reflection. Here, we need to understand only how they are conceptually motivated in a pragmatist spirit.

Our project recommends a seemingly indirect approach.

I

In explicating Leibniz's analysis of the nature of necessity, Alvin Plantinga introduces the notions of 'a possible state of affairs' and of 'a possible world' – since, as he remarks, 'we can do no better' (than Leibniz, here).[2] Roughly, by 'a possible state of affairs,' he says, we mean 'a *way things could have been*,' a logically compossible condition, something normally captured by or corresponding to a proposition – in the way in which (as Plantinga offers) the possible state of affairs *Socrates' being snubnosed* is captured by or corresponds to the proposition *Socrates is snubnosed*. Perhaps there are two distinct entities here, states of affairs and propositions; or perhaps, as Roderick Chisholm holds, there is only one kind of entity, states of affairs.[3] 'But not every possible state of affairs,' Plantinga goes on to explain, 'is a possible world. To claim that honor, a state of affairs must be *maximal* or *complete*' (which *Socrates' being snubnosed* is not); and 'a state of affairs S is complete or *maximal* if for every state of affairs S', S includes S' or S precludes S' ...; a possible world is simply a possible state of affairs that is maximal.'[4] Plantinga then leaps to the finding: 'Of course *the actual world* is one of the possible worlds; it is the maximal possible state of affairs that is actual, that has the distinction of actually obtaining.'[5]

But the least reflection shows that we do not know – and we do not know how to show – that whatever actual *states of affairs* obtain actually could function to define or characterize, *not* merely to designate, that possible world that *is* the actual *world*. By hypothesis, we already know that given states of affairs are actual states of affairs: they must, therefore, belong to the actual world; but we have no idea *how* to identify *that* actual state of affairs that *is the actual world* – as opposed, say, to identifying a possible (because actual) state of affairs that trivially belongs to, or is only a part of, or obtains in, the actual world. There is good reason to think that, in a cognitively serious sense, we *never* know what the actual world (the 'maximal possible state of affairs') is – we never know what the proposition is that, if true, would capture or correspond to the actual state of affairs that is the actual world; hence, we know only, in knowing what is actual, no more than that *that* is (for logically trivial reasons) part of, or obtains in, the actual world. In referring to the actual world, then, we refer to what is, to what we take to be, maximal or complete, but we never actually formulate the proposition corresponding to the determinate states of affairs in virtue of which that world *is* demonstrably maximal or complete.

By contrast, it is most instructive and quite suspicious to note that, in talking of possible but non-actual worlds, we cannot even get a grip on the demarcation between a possible world and a possible state of affairs: for instance, why could not *Socrates has an aquiline nose* serve to designate a possible world rather than merely a possible state of affairs within an as yet unspecified possible world? There seems to be no answer. Hence, contrary to what Plantinga apparently wishes to hold, the state of affairs *Socrates' being snubnosed* might, against all our protests, designate *a* possible world – and not merely a limited state of affairs within a possible world – *for some possible but non-actual world*. It might, in fact, do both – affecting, for that reason, robust comparison with the actual world. Actual – and merely possible – worlds-talk is simply not pertinently uniform. Regarding the first, we cannot fail to make reference to the actual world, though we have no idea of all that it praxically includes; given Plantinga's gloss, we cannot know what specifically is complete or maximal about it. Regarding the second, we cannot say how the constraints of being complete or maximal obtain, in virtue of which to sort out possible states of affairs within a possible world – as opposed to some specifiable possible world itself. This shows the peculiar vacuity of Plantinga's formula: 'A proposition is true in the actual world if it is true; it is true in [world] W if it would have been true had W been actual'; or that 'Each world W has the property of actuality in W (and nowhere else).'[6] This could have been said just as perspicuously without mentioning possible (non-actual) worlds at all. In fact, the salient difference between discourse about the actual world and about possible worlds is simply that, relative to the first, we suppose that much is included in that world that is *logically* independent of whatever else is included; and that, relative to the second, we have *no* clues as to what comparable constraints or tolerances should be imposed on what is included, apart from constraints of logical consistency alone. This is why it is idle to speak about what is true in a possible world W 'had W been actual.'

We may then sadly conclude that, apart from compossibility itself, with respect to the actual world, we have no idea how to proceed to show that what we *take* to be the actual world *is* maximal or complete; and with respect to possible worlds, we have no idea how logically to *encumber* the concepts of being maximal or complete, in order to distinguish between possible worlds and possible states of affairs within possible worlds. Of course, as we may suspect – and as Plantinga's own illustration seems to confirm – our notion of possible worlds is parasitic on our notion of the actual world, even though among those who use the possible-worlds idiom (Plantinga, in particular) the actual world is supposed to be individuated within the set of all possible worlds.

That we cannot actually individuate the actual world affects and complicates the import of Plantinga's account. For Plantinga goes on to say: 'Obviously *at least* one possible world obtains [that is, is actual]. Equally obviously, *at most* one obtains; for suppose two worlds W and W* both obtained. Since W and W* are distinct worlds, there will be some state of affairs S such that W includes S and W* precludes S. But then if both W and W* are actual, S both obtains and does not obtain; and this, as they say, is repugnant to the intellect.'[7] But Plantinga's conclusion presupposes that we are able to fix the actual world in some cognitively pertinent way: for if we do, we will (so he claims) be rationally obliged to concede that there is at least and at most the one actual world. He also holds that, 'It is the notion of existence *simpliciter* that is basic; existence-in-W is to be explained in terms of it.'[8] But he does not explain what it is to exist *simpliciter* or how we ever come to know that the snubnosed Socrates exists, in any sense conformable with his claim about the actual world. He considers that existence may be an 'atypical' property, but he does not develop the thesis in cognitively decisive ways.[9] Nevertheless, if the truth of propositions and the actuality of possible states of affairs are themselves inextricably linked with our conceptual schemes – in such a way that the world cannot be taken to be cognitively transparent, in such a way that no correspondence between true proposition and actual world can be established – then it is at least logically possible that Plantinga is simply wrong about the actual world.[10] In any case, Plantinga appears to set constraints on cognitively informed discourse about the actual world apart from considerations that are purely formal – but not restricted merely to formal coherence or consistency. But what we have shown is that reference to 'worlds,' either actual or possible, as opposed to reference to what is *in* 'a world,' plays no cognitively distinctive role beyond that of formal consistency.

In fact, there are at least two well-known, quite disparate ways in which Plantinga's claim has implicitly been challenged in recent years – both opposed to the Leibnizian option and both fairly characterized as pragmatist in motivation. W. V. Quine holds and has always held that, 'There is really only one world';[11] and yet, Quine's holism with respect to science entails an 'indeterminacy of translation' such that 'manuals for translating one language into another [featuring, of course, what science would favor as forming the most reliable core of its findings about the real world] can be set up in divergent ways, all compatible with the totality of speech dispositions, yet incompatible with one another.'[12] So, for Quine, the actual world is one, but what may count as true propositions and actual states of affairs is mediated by 'analytical hypotheses'[13] in such a way that incompatibilities among them *may* yet be tolerated as picturing or designating parts of the actual world (one as well as another) in spite of such distributed collisions. What this shows is that formal constraints on the compossibility of what is internal to postulated worlds, considered without regard to actuality, cannot as such prejudge what we may coherently claim to find with respect to what is actual. What we may claim regarding the actual world will inevitably be a function of how *we* characterize, of how we address, our own cognitive access to that world – a matter which cannot be convincingly managed in purely formal terms at all. Plantinga's claim, then, proves to be a straightforward *non sequitur:* Quine simply gives up the sanguine conviction that, apart from whatever is holistically confirmed regarding the actual world, we can fix distributive truths uniquely within the holistic body of science.[14]

The other alternative is the one favored by Nelson Goodman. Goodman begins with a consideration very much like Quine's – that is, with the rejection of cognitive transparency or correspondence.[15] But where Quine holds fast to one actual world, Goodman declares outright that, 'We are not speaking in terms of multiple possible alternatives to a single actual world but of multiple actual worlds.'[16] Many of course have been puzzled by this claim; but, conceding for the sake of the argument that it is coherent, Plantinga will be found to have characterized a possibility as 'repugnant to the intellect' that, on the evidence, need not be so at all. In any case, there surely *is* no way, from the vantage of Plantinga's stance, to decide in advance that Goodman's maneuver is an incoherent one – even if Goodman's specific version of the claim proves to be incoherent; also, as with Quine's option, it is quite possible that Goodman's option could be freed from whatever difficulties prove local to his own formulation. So we may risk going the extra mile with Goodman.

Goodman identifies himself as an 'irrealist,' by which he means that he is both 'an anti-realist and an anti-idealist'; co-ordinately, he is a 'radical relativist' with regard to these multiple actual worlds, in at least the double sense that he opposes 'linguistic universals' (binding on whatever language is alleged to constrain true descriptions of the world) and that he subscribes to 'ontological relativism' – which 'does not imply that all world-versions [as he terms his central notion] are right but only that at least some irreconcilable versions are right.'[17] Also, since, for Goodman, 'there is nothing nonactual,' possible worlds are disallowed; but, 'Since there are conflicting truths, there are many worlds if any, but no such thing as *the* world.' So Goodman rejects correspondence, true representation, or any 'faithful reporting on "the real world." For there is ... no such thing as the real world, no unique, ready-made, absolute reality apart from and independent of all versions and visions. Rather [he says], there are many right world-versions, some of them irreconcilable with others; and thus there are many worlds if any. A version [he adds] is not so much made right by a world as a world is made by a right version.'[18] Apparently, a 'world' is what we may term the internal accusative of a 'version': hence, Goodman, not altogether unlike Plantinga, must – if he is to individuate worlds – individuate versions; but he nowhere attempts to do so. Hence, he has shown neither how to preclude Quine-like 'incompatibles' relative to a *given* 'world-version' nor (as yet at least) what cognitively pertinent grounds we have or could have by reference to which particular apparently 'conflicting truths' could be shown to be falsehoods relative to *various* 'actual worlds.'

Up to this point the option looks coherent – though not entirely explicit – since the conceptual relation between 'versions' and 'worlds' and the meaning of 'world-versions' are unclear. On this provisional reading, Goodman sketches an option according to which Plantinga is multiply mistaken: for (a) if there is at least one actual world, there are many actual worlds; (b) there is no sense in which actual worlds are members of the set of possible worlds (including non-actual worlds); and (c) inasmuch as multiple actual worlds include ('irreconcilable' worlds, there are or may be 'conflicting truths' with regard to what is actual. The point again is that the option arises with respect to what we may say about human conceptualization and knowledge, *not* merely about formal constraints on consistency.

What is troublesome in a sympathetic way about Goodman's account is that he fails to address what is critical to his own philosophical program. For, for one thing, he does not tell us fully how to distinguish between competing ('irreconcilable') theories with regard to, or falling within, one and the same 'world' or 'world-version' and competing ('irreconcilable') 'world-versions': this, as already remarked, returns us to a difficulty regarding what is 'maximal' and 'complete' in Plantinga's account. And for another, he does not tell us fully what the relationship is between 'version' and 'world': and this, of course, affects what we should count as true among propositions, and 'right' among would-be 'world-versions.' On Goodman's view, consistently with his irrealism, there cannot be any privileged vantage from which to determine which world-versions are 'right' or which detailed propositions – about whatever falls within given world-versions – are 'true.' As he says, 'There is no version-independent feature, no true version compatible with all true versions.' Consequently, 'Everything including individuals is an artefact.' Nevertheless, 'Right versions are different from wrong versions.' How can this be? The answer Goodman gives is this: 'If we make worlds, the meaning of truth lies not in these worlds but in ourselves – or better, in our versions and what we do with them.'[19] But if *we* ourselves are not artifactually made within particular world-versions but *make* them instead, then it is difficult to see how Goodman can escape some part of the cognitive privilege he denies; and if we *are* artifactually within them, then it is difficult to grasp what it would mean to claim a distinction between right and wrong world-versions.

The closest Goodman comes to providing an answer returns us to his theory of projectibility at least as an illustration of the answer he apparently has not yet completely formulated. To the question, 'What makes a category right?,' he answers: 'Very briefly, and oversimply, its adoption in inductive practice, its entrenchment, resulting from inertia modified by invention.' So inductive validity is at least 'an example of rightness other than truth'; and, although 'we cannot equate truth with acceptability' (since 'we take truth to be constant while acceptability is transient'), nevertheless '*ultimate* acceptability – acceptability that is not subsequently lost – is of course as steadfast as truth.' In this sense, 'Although we may seldom if ever know when or whether it has been or will be achieved, [ultimate acceptability] serves as a sufficient condition for truth.'[20]

But this will not do at all, since the rather Peircean long run ('entrenchment') that Goodman appears to count on would *require one* actual world with respect to which our versions are fallibilistically and progressively linked, would *require* a world that is not *merely* an artifact of our conceptual world-versions, would *require* a universal community of inquiring minds functioning self-correctively *with respect to* the one world to which they belong. If he does not count on that, we should be even more baffled about the 'right' constraints to be imposed on the plural actual worlds we make. In other words, *if* Goodman's thesis is coherent, it must account for the *non-artifactual* aspects of what is actual – consistently with irrealism, relativism, and a restricted constructivism. It would still be different from Quine's view, but it would not address the *cognitive* concern that it means to espouse. Curiously, then, there is almost nothing in either Quine's or Goodman's account that directly addresses the conditions of actual human inquiry or cognition, any more than there is in Plantinga's – which confirms the vacuity or near-vacuity of talking about 'the' one world or about many 'worlds,' actual or possible.

What is instructive about our survey is that we see that constraints on our discourse about what is actual – whether in accord with alternative Leibnizian[21] or anti-Leibnizian idioms – *cannot* be pertinently motivated in purely formal terms, *either* in modal terms *or* in terms addressed to the conditions of a human science. Quine's and Goodman's efforts signify the open-ended nature of theorizing about the conditions of empirical truth and intelligibility – that the Leibnizian cannot preclude or relevantly guess at. But for their own part, Quine and Goodman themselves are content merely to specify what certain global options are *if* we reject the cognitive transparency of the world. They have no theory of the actual behavior of scientific thinking considered historically or diachronically. That is the trouble, for instance, with Goodman's notion of 'entrenchment' – particularly with his notions of 'earned' and 'inherited' entrenchment and with his notion of 'differences in *degree* of projectibility' – since, under historicized and radically relativized circumstances, the governing notion of '*ultimate* acceptability' and its regularized bearing on these other distinctions are rendered completely meaningless or inoperable.[22] Similarly, when Quine summarizes 'the genesis and development of reference' applied 'to either the individual or the race,' when he supplies what he himself terms the 'psychogenesis' of reference, he means only to have provided an *imaginary* reconstruction of how, in natural language, we *could* have developed the practice of 'objectual' quantification.[23] There is no account in either Plantinga or Quine or Goodman of our conceptual resources considered as part of the historical development of human inquiry. In this sense, Quine and Goodman may be termed *formal pragmatists,* since, in rejecting the cognitive transparency of the world or worlds, they conform minimally to what, both historically and as a useful term of art, is required by a pragmatist view of science.[24] Also, in this regard, they gain a march on Plantinga, since the strongest arguments drawn from the most varied sources of current philosophy appear to reject transparency and a cognitive reading of the correspondence theory. Still, on the argument offered, it remains quite impossible to decide, with Goodman, between right and wrong 'world-versions,' or, with Quine, between competing hypotheses under the conditions of a thoroughgoing scientific holism, unless we can legitimate our reliance on particular historically emergent and contingently favored schemes for characterizing and testing the presence of what we take to be actual phenomena. There is a remarkably large *lacuna,* therefore, in these thin pragmatisms.[25]

II

Viewed against this backdrop, the distinction of Martin Heidegger's *Being and Time* lies primarily in Heidegger's having attempted a comprehensive phenomenology of coming to know the things of the world under the most radical reading of the pragmatist denial of cognitive transparency. This Heidegger achieves partly by characterizing man as *Dasein:* for, by this designation, Heidegger means to signify that man is essentially a historical or historicized being; that being such is the tacitly predisposing precondition for whatever in his temporal experience he may be said to come to know; that the things of the world and their natures are grasped by man (*Dasein*) in his existential or continually present activity in the world; and, most important, that in being *Dasein,* the entity, man is essentially defined by the power through which the structure of things is 'disclosed' to him (serially, contingently, without privilege, under prejudice), through his own interested activity, subject to the blind forestructuring of his cognitive and active capacity. Heidegger achieves his picture of man's cognitive role by identifying the objects thus disclosed (the 'phenomena') as objects whose disclosed structure is a structure that itself depends on what can only be specified mythically as the 'relationship' between *Dasein* and *Sein*. Such objects and their structures are constituted *in* being disclosed to *Dasein:* the cognizing self (the 'existentiell' self) is included among such 'objects.' This is what Heidegger means by viewing matters phenomenologically.[26] In that process, *Dasein* reflexively comes to understand itself.

The following may be offered as a fairly perspicuous statement of Heidegger's own summary view, if we are prepared to tolerate the alien quality of Heidegger's prose as seen from the vantage of analytic philosophy:

Dasein [he says] always understands itself in terms of its existence – in terms of a possibility of itself; to be itself or not itself. Dasein has either chosen these possibilities itself, or got itself into them, or grown up in them already [and] the understanding of oneself which leads *along this way* we call '*existentiell.*' ... So far as existence is the determining character of Dasein, the ontological analytic of this entity always requires that existentiality be considered beforehand. By 'existentiality' we understand the state of Being that is constitutive for those entities that exist. But in the idea of such a constitutive [or ontic] state of Being, the idea of Being is already included [ontologically, not merely ontically]. ... Sciences are ways of Being in which Dasein comports itself towards entities which it need not be itself. But to Dasein, Being in a world is something that belongs essentially. Thus Dasein's understanding of Being pertains with equal primordiality both to an understanding of something like a 'world,' and to the understanding of the Being of those entities which become accessible within the world. So whenever an ontology takes for its theme entities whose character of Being is other than that of Dasein, it has its own foundation and motivation in Dasein's own ontical structure [that is, in whatever of its own structure is disclosed in the ongoing *existentiell* way].[27]

This is certainly not what Quine or Goodman would ever say. But it is, correctly read, a way of historicizing what, in the most radical sense, is common to the denial of the cognitive transparency of the world – in Quine's holism, Goodman's irrealism, and Heidegger's existentialized phenomenology. It is a way of preserving the sense of man's effective survival in the world (a sense much more robust than either Quine's or Goodman's; more like Peirce's or James's or Dewey's). In effect, Heidegger identifies the minimal conditions and preconditions of human knowledge a *pragmatist* conception of science requires; his analysis fixes the sense in which whatever structures science imputes to the world depend entirely upon and are formulable only within the historical processes of *praxis* and technology. In effect, Heidegger *legitimates* science technologically *and* holistically; in doing that, he probes what is ontologically (or phenomenologically) *prior* in *man's being capable of* cognitive discovery within-the-world – the (necessarily mythologized) precondition on which disputes about distributed claims in science must depend.

Speaking this way may still obscure the peculiar power – and limitation – of Heidegger's contribution. A dialectical comparison between the formal theories offered by Quine and Goodman and Plantinga's refinement of a Leibnizian formalism shows: (i) that would-be constraints of consistency and coherence imposed on our theory of the actual world *cannot* be given logical priority over our detailed interpretation of the world and our cognitive powers; formalism is simply pointless, forever generates question-begging 'necessities'; (ii) that reference to a 'complete' and 'maximal' world or worlds, actual or possible, never plays a *cognitively* productive role in developing a science; the individuation of worlds is conceptually idle; and (iii) that the denial of the world's cognitive transparency affords a direct conceptual basis for viably accommodating the *permanent* presence of incompatible sets of truth claims; alethic questions no longer take precedence over epistemic questions. So there is an enormous gain to be made in moving beyond a Leibnizian formalism to a pragmatized formalism, to a mythic phenomenology drawn from the technologized preconditions of inquiry. Nevertheless, that gain is also a dead end if it is not supplemented internally by a further account of an effective and detailed theory of science. Neither Plantinga nor Quine nor Goodman nor Heidegger addresses the issue.

Heidegger's thesis fills a gap, then, by specifying an existential precondition affecting man's (or *Dasein's*) *cognitive* power in two distinct but inseparable ways: first, in that particular cognitive claims are, necessarily, instantitations of *Dasein's* peculiar mode of being *vis-à-vis* whatever of reality may prove to be cognitively disclosable; and second, in that the conceptual structure of particular such claims is always preconditioned by *Dasein's* already having been preconditioned by its own cognitive (or 'existentiell') history and by its having recovered that history under the same condition.

Technology is Heidegger's master clue: not in the sense that cognitive discovery is exclusively technological but in the sense that disclosure through human technology is always primary, salient, effectively linked to what may be disclosed even if by another avenue of *Dasein's* capacity (poetry or art, for instance).[28] Technology draws attention to the inchoate sources within human existence itself – on which human viability depends – in virtue of which the world is seen to be aptly structured *for* cognitively informed activities of any kind; it is also drawn to the power of discontinuous, contingent, serial, variably overriding or influential such disclosures that continually reorient cognition and science historically and reveal to man his own profoundly historical nature. In this sense, Heidegger may not unreasonably be said to provide a phenomenological basis for the more anthropological theses advanced both in Thomas Kuhn's account of normal and revolutionary science and Michel Foucault's account of the unpredictable seriality of the conceptual orientations of different historical ages.[29] But they, unlike Heidegger, actually address detailed first-order claims.

There is no point to the phenomenological thesis about technology unless it is closely linked with an actual account (at the level of psychology, ideology, history, sociology, hermeneutics, critique – or of a more generalized philosophical anthropology) of how the 'technic' conditions the actual horizon of determinate inquiry; on the other hand, there is a kind of systematic scatter in a collection of such accounts if we fail to relate their shifting conceptual tableaux to the conditions of man's existence. *There* is the point of featuring the peculiarly abstract formulations of Quine's holism and Goodman's irrealism. It is apparent to any reader of *Word and Object* that Quine never addresses the technological – or pragmatist – aspects of the mere survival of a human society, in virtue of which, say, a field linguist seeks to penetrate an alien language. It is surely only for this reason that Quine theorizes that the 'analytical hypotheses,' by which we parse a would-be language, form a perhaps inescapable but nevertheless arbitrary imposition of one human society's conceptual orientation on another's. It is similarly apparent to any reader of *Fact, Fiction, and Forecast* and *Ways of Worldmaking* that Goodman never addresses the detailed practices of actual inquiry in terms of which to assess the projectibility of competing categories of induction. The unexpected insight regarding Quine and Goodman that Heidegger's approach to the question of science provides is simply that their resolution cannot be derived from timeless or essentialist principles (with which, of course Quine and Goodman would 'formally' agree) but is, withal, existential, plausibly imputed (not uniquely and not self-evidently) by attention to the particular technological conditions under which given historical phases of human inquiry are seen to be ordered.

The Heideggerean view – which we are here rather unabashedly integrating with American pragmatism – may be conveniently fixed by a few brief citations and allusions. First of all, from the phenomenological point of view, we are to understand that what we come to know and treat as distinct worldly things (objects or phenomena) we do treat thus – 'we encounter as closest to us' (as Heidegger puts it) – because of our 'dealings' (*Umgang*) '*in* the world and *with* entities within the world.'[30] That encounter is, then, not yet cognitive but the precondition of our being cognitively concerned with whatever is thus disclosed. The most reasonable interpretation of the thesis is that, at the very least, man's mode of existing (including the subterranean or tacit capacity for survival) skews or forms or founds whatever appear as objects-discovered-by-cognizing-subjects. Here, Heidegger borrows the Greek term for things – *prágmata* – 'that which one has to do with in one's concernful dealings (*praxis*).' But the Greeks (he claims) failed to explicate what such things were: they thought of them ' "proximally" as "mere things" [and hence, were deceived by their metaphysics]. We [Heidegger adds] shall call those entities which we [thus] encounter in concern "*equipment*" [*das Zeug*].'[31]

From Heidegger's point of view, the categorizing of things cognized is, in some sense, originally generated by reflecting on what we emergently find suited to our existential dealings – suited to survival, in particular. Heidegger hastens to add here: 'Taken strictly, there "is" no such thing as *an* equipment. ... Equipment is essentially "something in-order-to..." [*etwas um-zu...*]. A totality of equipment is constituted by various ways of the "in-order-to," such as serviceability, conductiveness, usability, manipulability.'[32] This captures the complex idea that our classificatory schemes are not informed by the cognitive transparency of the actual world (the correspondence thesis); that the classificatory schemes we favor reflexively emerge from our existential dealings with the world in which we find ourselves (the phenomenological aspect of technology); and that these classicatory schemes have an historically contingent life, gradually replacing or affecting one another, legitimately within but only within the interval in which they thus function (the very meaning of *Dasein* itself). This is the reason for linking Heidegger with the essential themes of American pragmatism: Heidegger says that 'Nature is not to be understood as that which is just present-at-hand [*Vorhandenheit*], nor as the *power of Nature.*' No, 'the ready-to-hand [*Zuhandenheit*] is not thereby *observed* and stared at as something present-at-hand; the presence-at-hand which makes itself known is still bound up in the readiness-to-hand of equipment.'[33]

Shorn of its somewhat picturesque terminology, what Heidegger is getting at is that the apparent, cognizable structures of objective nature or reality are originally founded upon the existentially historicized condition of human life itself, which – viewed as the precondition of our science – generates our shifting picture of the known world as an abstraction from what reflexively we must suppose to have been disclosed in our technologized mode of surviving. We are gifted enough to survive in a world in which we 'deal' with 'things' before we know what they are – and in which, therefore, we identify things as what they are (objectively) because of that contingent and changing linkage. It would be too weak to say (in Heidegger's sense) that things are, then, first relationally identified in terms of their use as gear or equipment. No, in existing, in surviving, *we* reflexively see that we must have first tacitly *used* things *as* equipment. Our various classificatory schemes derive from *that* condition – *but not cognitively at the point at which that condition (mythically) obtains.*

Here, therefore, Heidegger seeks to go beyond the philosophical anthropology implicit in Kant's first *Critique* to the phenomenological source of all such anthropologies. That theme, the theme that our cognitive abilities *cannot be cognitively grounded,* is the *sine qua non* for effectively rejecting foundationalism, the philosophy of presence, the correspondence theory, and the like. Formulated in terms of a realist science, the theme is also the central theme of pragmatism. Thus, the seemingly baffling puzzles of Quine's and Goodman's invention depend upon – accordingly, are also resolved by – attending to the *supposed* need – now impossible to satisfy – of grounding questions of objective truth in some account of the world's assured transparency. Quine's and Goodman's paradoxes only arise on a theory that they themselves disallow. *They* oblige us to give up that theory all right, but they do not withdraw the associated expectations. Hence, the apparent bite of the indeterminacy of translation and of the conflict of incompatible but actual plural worlds. Heidegger's thesis about technology, then, offers a felicitous way of rejecting the transparency doctrine and of dissolving the puzzlement that still adheres because we *once* subscribed to it. Quine's and Goodman's worries are therefore vestigial.

III

Finally, let us risk a brief comparison between Heidegger and Dewey. The following is tantalizingly close (however alien) to Heidegger's idiom:

> By its nature [says Dewey] technology is concerned with things and acts in their instrumentalities, not in their immediacies. Objects and events figure in work not as fulfillments, realizations, but in behalf of other things of which they are means and predictive signs. A tool is a particular thing, but it is more than a particular thing, since it is a thing in which a connection, a sequential bond of nature is embodied. It possesses an objective relation as its own defining property. ... Man's bias towards himself easily leads him to think of a tool solely in relation to himself, to his hand and eyes, but its primary relationship is toward other external things, as the hammer to the nail, and the plow to the soil. Only through this objective bond does it sustain relation to man himself and his activities. A tool denotes a perception and acknowledgement of sequential bonds in nature.[34]

Notice that Dewey's notion of the 'primary relationship ... toward other things' corresponds to Heidegger's 'totality of equipment' – *not* to what Heidegger terms 'presence-*at*-hand' (*Vorhandenheit*). Notice, also, that man's reflexive understanding is mediated by technologized activity and perception. The entire discussion, in fact, is set in terms of Dewey's opposition to all the dualisms he believed he could draw from the classic philosophy of the West. There is no question that Dewey was not attracted to anything like Heidegger's high-blown phenomenology. In this sense, he was perhaps more inclined toward a philosophical anthropology in the sense already suggested. Nevertheless, one can also find in Dewey remarks like the following:

> But if we free ourselves from preconceptions, applications of 'science' means application *in*, not application *to*. Application *in* something signifies a more extensive interaction of natural events with one another, an elimination of distance and obstacles; provision of opportunities for interactions that reveal potentialities previously hidden and that bring into existence new histories with new initiations and endings.[35]

This is an extraordinarily close approximation, within the idiom of American pragmatism, of Heidegger's contrast between *Vorhandenheit* and

Zuhandenheit, with a splash of the existential history of *Dasein*. The point, however, is not to claim an identity of positions, only a strongly convergent line of theorizing that promises the most balanced and most comprehensive account of the full meaning of technology. Furthermore, on the argument here developed, although there is a difference between the inquiries of a philosophical anthropology (or a transcendental philosophy) and a phenomenology of man, there is no difference that would legitimate a claim about the relative privilege or foundational nature of the one inquiry over the other. We may distinguish first- and second-order inquiries for convenience; but to treat technology as a theory of man's historicized existence is to deny the cognitive ascendency of first- and second-order inquiries in either direction. Also, it must be said, the legitimated transcendental philosophy is *not* a substitute for any distributive, first-order science or for any first-order critique (whether Marxist, Frankfurt-Critical, Freudian, or Foucaultian) of the governing prejudice, interest, ideology, horizon, or the like, with which it is effectively pursued.

Temple University

NOTES

[1] I have explored this convergence in 'Entailments from "Naturalism-Phenomenology,"' forthcoming in a collection edited by Mark Amadeus Notturno.
[2] Alvin Plantinga, *The Nature of Necessity* (Oxford: Clarendon, 1974), p. 44.
[3] Roderick M. Chisholm, 'Events and Propositions,' *Noûs*, 4 (1970); 'State of Affairs Again,' *Noûs*, 5 (1971). These are cited by Plantinga.
[4] *Op. cit.*, pp. 44–45.
[5] *Ibid.*, p. 45.
[6] *Ibid.*, pp. 46–48.
[7] *Ibid.*, p. 45.
[8] *Ibid.*, p. 46–47.
[9] *Ibid.*, p. 137; see also Alvin Plantinga, *God and Other Minds* (Ithaca: Cornell University Press, 1967), chapter 2. See, further, Joseph Margolis, *Knowledge and Existence* (New York: Oxford University Press, 1973), chapter 4.
[10] The most fashionable survey of the current tendency to reject cognitive transparency may be found in Richard Rorty, *Philosophy and the Mirror of Nature* (Princeton: Princeton University Press, 1979).
[11] W. V. Quine, *Methods of Logic* (New York: Holt, 1950), p. 200; see also W. V. Quine, 'Goodman's Ways of Worldmaking,' *Theories and Things* (Cambridge: Harvard University Press, 1981).
[12] W. V. Quine, *Word and Object* (Cambridge: MIT Press, 1960), p. 27.
[13] *Ibid.*, § 15.

[14] *Ibid.*, § 56. (There is no question, it should be added, that the thesis needs some adjustment in terms of Quine's actual statements, because Quine's formulation may well be incoherent as it stands. But it does not seem to be critical to the possibility at stake.) See, further, Joseph Margolis, 'The Locus of Coherence,' *Linguistics and Philosophy*, 7 (1984).

[15] Nelson Goodman, *Of Mind and Other Matters* (Cambridge: Harvard University Press, 1984), chapter 2.

[16] Nelson Goodman, *Ways of Worldmaking* (Indianapolis: Hackett, 1978), p. 2.

[17] *Of Mind and Other Matters*, pp. 7, 13.

[18] *Ibid.*, pp. 125, 127.

[19] *Ibid.*, pp. 29, 33, 38. In his brief review of the book, Quine appears to have found this line of thinking either incoherent or hopelessly extravagant. See also Hilary Putnam, 'Reflections on Goodman's *Ways of Worldmaking*,' *Journal of Philosophy*, 76 (1979); Israel Scheffler, 'The Wonderful Worlds of Goodman,' *Synthese*, 45 (1980); C. G. Hempel, 'Comments on Goodman's *Ways of Worldmaking*,' *Synthese*, 45 (1980).

[20] *Ibid.*, p. 38.

[21] For an alternative to Plantinga's view of possible worlds, see for instance David Lewis, 'Counterpart Theory and Quantified Modal Logic,' *Journal of Philosophy*, 65 (1968); 'Anselm and Actuality,' *Noûs*, 4 (1970).

[22] Nelson Goodman, *Fact, Fiction, and Forecast* (2d ed.; Indianapolis: Bobbs-Merrill, 1965), pp. 105–106.

[23] W. V. Quine, *The Roots of Reference* (La Salle: Open Court, 1973), p. 123.

[24] See Joseph Margolis, 'Pragmatism without Foundations,' *American Philosophical Quarterly*, 21 (1984); this has been incorporated in *Pragmatism without Foundations* (Oxford: Basil Blackwell, 1986).

[25] The term 'pragmatism' is not inapt as applied to Goodman's position, in spite of the fact that Goodman had, earlier on, indicated his at least partial opposition to pragmatism. For Goodman had, in developing his theory of projectibility with regard to both undetermined cases and truth, maintained that, 'Since a hypothesis is true only if true for all its cases, it is true only if true for all its future and all its undetermined cases; but equally, it is true only if true for all its past and all its determined cases,' *Fact, Fiction, and Forecast*, p. 91, note 3; this must be read in the context of pp. 89–99. But now, conceding 'conflicting truths,' it is no longer clear that the required entrenchment of genuine projectibles can be measured or measured with respect to 'ultimate acceptability.'

[26] Martin Heidegger, *Being and Time*, trans. from 7th German ed. John Macquarrie and Edward Robinson (New York: Harper and Row, 1962), p. 50. What is offered here, thus far, is of course an extremely abbreviated summary of a sprawling theme. But it is, perhaps not unfairly, offered as the nerve of the Introduction.

[27] *Ibid.*, p. 33; see p. 34.

[28] Martin Heidegger, 'The Origin of the Work of Art,' in *Poetry, Language, Thought*, trans. Albert Hofstadter (New York: Harper and Row, 1971).

[29] See Thomas S. Kuhn, *The Structure of Scientific Revolutions* (2d ed.; Chicago: University of Chicago Press, 1970); and Michel Foucault, *Les Mots et les choses*, translated as *The Order of Things* (New York: Random House, 1970).

[30] *Being and Time*, p. 95.

[31] *Ibid.*, pp. 96–97.
[32] *Ibid.*, p. 97.
[33] *Ibid.*, pp. 100, 104.
[34] John Dewey, *Experience and Nature* (2d ed.; New York: Dover, 1958), pp. 122–123.
[35] *Ibid.*, p. 162.

PART III

INTERNATIONAL AND INTERGENERATIONAL PERSPECTIVES

GAO DASHENG AND ZOU TSING

PHILOSOPHY OF TECHNOLOGY IN CHINA

1. THE RISE OF PHILOSOPHY OF TECHNOLOGY IN CHINA

In China, just as in many Western countries, philosophy of technology has only recently become a branch of philosophy. Before this, though Chinese Marxist philosophers devoted considerable energy to discussing traditional Marxist problems related to technology – e.g., the interaction of the forces of production and the relations of production – none took technology as an independent object for philosophical discussion.

Since the end of the 'Great Cultural Revolution' (1976), however, China has been confronted by the important problem of raising the level of science and technology as fast as possible. In this social setting, Chinese scholars gradually have become interested in such problems as the nature of science and technology, their position and role in society, and the pattern of their development. Philosophy and sociology of science, management of research and development, science and technology policy studies, and history of science and technology have occupied unexpectedly important positions in the sphere of learning since 1976. Shortly thereafter research concerning philosophy of technology also began to flourish.

Studies of the history of technology in China in which a few scholars considered technology from a philosophical perspective began fairly early. The Committee for the History of Technology was set up after the Chinese Society for the History of Science and Technology resumed its academic activities. The First National Congress of the History of Technology was held in Wuhan in 1979. Subsequent Congresses were held regularly in 1981, 1984, and 1986; at each congress many scholars discussed problems concerning philosophy of technology, especially the definition of technology and the patterns or laws of technological development.

In 1979, in order to develop undergraduate science, technology, and society (STS) education, the Ministry of Education organized a number of experts to write a textbook, *Teaching Materials on Natural Dialectics,* and a series of monographs, one of which, *Some Dialectic Contents in Engineering and Technology,* was written by scholars of the Middle China Institute of Technology. This first monograph in the field of philosophy of technology in

China has had a great impact on later Chinese studies of philosophy of technology.

After this, the literature discussing technology as a philosophical issue began to increase. In 1982, Yuan Deyu's essay, 'Technology as an Independent Issue of Study,' which suggests that technology must be studied as a philosophical subject in its own right, was published in the *Newsletter on Natural Dialectics*. Huang Linchu's essay, 'Problems Studied by the Philosophy of Technology,' which enumerates a series of questions, was published in the same journal in 1982. *Information from Nature*, a journal which started publication in 1980, gradually became an important outlet for academic discussion on the philosophy of technology. Almost every issue has published two to four papers on related topics.

Other journals have also published many articles about philosophy of technology. In 1984, *Philosophical Research*, the leading journal of philosophy in China, initiated a special section, 'The New Technological Revolution and the Development of Society,' that has included some dozen articles and provoked much controversy – controversy which has not yet to subside. In 1985 the *Journal of Dialectics of Nature*, the leading journal on philosophy and sociology of science, published Chen Changshu's 'Technology as an Object of Study for Philosophy,' in which Chen outlines twelve questions to be dealt with by any philosophy of technology. Among these are:

> What is technology?
> What essential factors constitute technology?
> How is technology to be classified?
> What is the impact of technology on society?
> What are the relations between technology and other social factors (including social ideology)?

After three years of planning and preparation, the First National Congress on the Philosophy of Technology, sponsored by the Chinese Society for Dialectics of Nature, was held in Chengdu, November 12–19, 1985. The concepts of essence, essential factors, and structure of technology – the so-called 'problems of the ontology of technology' – were the central issues in the congress. Other problems, such as those of the relation between technology and society, were touched on as well. At the end of the congress, a Committee for the Philosophy of Technology was set up to coordinate and organize academic activities in all parts of the country and to arrange for a Second National Congress of the Philosophy of Technology in 1988.

In China, studies in philosophy of technology have developed in a few universities, especially in institutes of technology. Some scholars in the Northeast Institute of Technology have devoted considerable attention to the study of Japanese philosophy of technology, introducing it into Chinese academic circles. The first research group on the philosophy of technology in China was set up in Tsinghua University in February, 1987. This group is working to promote the development of Chinese philosophy of technology through the assimilation of the achievements of scholars in other countries.

The meaning of 'philosophy of technology' in Chinese is not exactly the same as in English. In Chinese, 'philosophy of technology' is called 'jishu luen.' 'Jishu' means 'technology,' and 'luen' means that a certain object is taken as a theme for study, although not necessarily that this study is exclusively philosophical. Though 'luen' is translated into English as 'philosophy,' the Chinese word whose meaning is closest to this English term is actually 'zexue.' Some Chinese studies of philosophy of technology exhibit a pronounced practical emphasis, which resembles the sociology of technology more than the philosophy of technology. This essay, however, will stress those studies which have a strongly philosophical cast in the English sense.

2. THE MAIN PROBLEMS IN CHINESE PHILOSOPHY OF TECHNOLOGY

The main problems discussed in Chinese philosophy of technology can be classified under five headings: (1) those dealing with the essence and essential factors of technology; (2) those focusing on the system of technology and the pattern of its development; (3) those concerning the relationship between technology and society; (4) those stressing issues of the relationship between technology and nature; and (5) those analyzing the methodology of technology. Not all Chinese scholars, however, would agree with this classification; also, this classification is not exhaustive of the whole of Chinese philosophy of technology. Neither are these five sets of problems wholly independent of one another; it is difficult to subsume some studies within any aspect of this simple framework or, perhaps, any other framework.

The Essence and Essential Factors of Technology

Philosophy of technology as a developing discipline must delimit its object of study. When technology is taken as an object of philosophical investigation, questions dealing with its essence or essential characteristics – those features which distinguish it from other things such as science – naturally become the first thing to be discussed.

Definitions of technology provided by Japanese scholars have been the starting point of Chinese philosophical research. Japanese philosophy of technology has discussed three types of definition; (a) technology is a kind of human capability; (b) technology is a kind of knowledge; (c) technology is a system of material means to a certain end. Yan Kangnian holds that the second is correct. In his 'Inquiry into Some Fundamental Problems in the Study of the Theory of Technology,' Yan defines technology as 'systematic knowledge and skill about making and operating.'

Most Chinese scholars take a more circumspect approach to the definition of technology. Yuan Deyu and Chen Changshu (1986) argue that all three kinds of definitions are one-sided. They regard technology as 'a system composed of all essential factors which are interrelated, or a dynamic system or process.' But they also think that considering technology as a system or dynamic process indicates not so much a definition as a crucial change in approach to technology – from a static, analytical investigation to dynamic, comprehensive research. The static, analytical approach to technology as action, human capability, or objective material substance cannot, in Yuan's and Chen's view, reveal the essence of technology. A conception of technology, as Zou Shangang (1985) points out:

is the starting point and end-result of a philosophy of technology. ... The wealth of information and possible contradictions included in it are enough to expand into an integrated philosophy of technology.

Thus, although it is necessary to investigate the definition of technology, it is unrealistic to desire an explicit and comprehensive definition at the outset of a philosophy of technology.

Another approach to technology is investigation into its essential factors. Zou Shangang and others, in their *Some Dialectic Contents in Engineering and Technology,* argue that to understand the law of development by internal contradictions, the dialectic essence in engineering and technology, it is necessary to distinguish the crucial and general factors, the essential factors of engineering and technology, from the numerous concrete relationships and factors that make up engineering and technology, and exist in the development of technology and engineering. These authors regard materials, energy, control, and craft as essential factors in engineering and technology; they discuss such problems as the characteristics of the essential factors, the relations between these, and how they promote technological change.

Zou's book has evoked considerable controversy. It focuses on whether or not materials, energy, control, and craft are essential factors. The answer to this question depends, however, on ideas about the essence of technology. For Yan Kangnian, the essential factors of technology are theory, design, craft, technological praxis, and skill, because he considers technology a kind of knowledge and skill. For Huang Linchu, there are two kinds of essential factors. One is material, another immaterial, because there are two kinds of technology – production technology and nonproductive technology. Regarding technology as a dynamic process, Yuan and Chen do not agree that materials are an essential factor because 'they do not take part in transforming nature directly,' and they maintain a cautious attitude toward whether or not energy and craft are essential, stressing that the question of essential factors must be investigated dynamically and genetically.

Yuan's and Chen's systematic and genetic approach is further developed in Chen Fan's 'On the Essential Factors of Technology and Its Structure.' Chen argues that the essential factors of technology constitute a system, so cause-and-effect mechanisms and the relationship between essential factors must be studied from a systematic perspective. The system of technology, in Chen's view, is an evolutionary one. The ancient system of technology was 'a one-dimensional system' that only contained one essential factor, a subjective one – namely, the experience and skill of the craftsman. With the coming of the Industrial Revolution, instruments and machines – objective factors – took a more and more important place in the system of technology. So, subjective and objective essential factors constitute the modern two-dimensional system of technology. In the contemporary system of technology, technological theory – applied scientific theory – plays a more important role than before. This is a new essential factor – in Chen's term, 'an objectified subjective essential factor of technology' – that, along with the two other essential factors, constitute the contemporary three-dimensional system of technology.

Chen uses a mathematical method to analyze the system of technology and its essential factors. Taking S as a system of technology with E as essential factors, the relations between technological system and essential factors can be shown in a set of differential equations:

$$dE_i/dt = S_i(E_1, E_2, E_3) \quad \text{where} \quad i = 1, 2, 3, \ldots n.$$

From such equations, Chen deduces four properties of the technological system.

The controversy concerning the essence and essential factors of technology is quite heated. The present review of this discussion has only introduced the most influential positions. Currently, however, literature on this topic tends to be gradually decreasing, although the conflict has not so much disappeared as spread into other areas, especially into questions related to the pattern of technological development.

The System of Technology and the Pattern of Its Development

In the development of the technological system, different kinds of technology restrict and interpenetrate each other, becoming eventually a whole. Research on the technological system aims to describe the ways this happens.

When approaching technology as a system, Chinese scholars largely agree on its characteristics. Liu Dongzhen and Yang Derong in their essay, 'On the Technological System,' argue that 'the technological system is an organic whole constituted of all kinds of technology aiming at achieving certain social ends and held together by internal mechanisms.' They think, like Yuan Deyu and Chen Changshu (1986), that there are three ways in which various technologies are linked with each other: (a) dependently, as when the emergence of technology X is a precondition for technology Y; (b) through infiltration, as when electric machines are adopted by numerous trades and professions; and (c) by chain relationships, as when the rise of one technology causes the emergence of many other technologies. Yuan and Chen also argue, as already mentioned, that not static but dynamic analysis is the proper way to reveal the system and structure of technology.

The pattern or law of technological development is a central problem in Chinese philosophy of technology. Research is so manifold on this topic that only a few influential ideas can be mentioned here.

In his essay, 'A Theory concerning the Systematic Evolution of Technology,' Lu Pinyue uses systems theory to analyze the structure of technology. He proposes three laws of technological evolution: (a) the law of 'revolution from quantitative change,' which specifies that the accumulation of technological change will to a certain extent bring about disruptive alterations in a technological paradigm; (b) the law of vertical action, in which a technological revolution at one level brings about a technological improvement at another; (c) the law of the horizontal sequence of technological paradigms. So far there have been three technological paradigms in history: stone implements and rope → machine technology; using of fire → chemical technology; farming, animal husbandry, and medicine → biological technology.

In his essay in *The Principle of Natural Dialectics,* Huang Linchu considers problems dealing with the formation of the technological system and the pattern and trends in its development. He describes the basic modes of technological development as including invention, the development of new technological principles, technological improvement, technological revolution and innovation, technological transfer, and he analyzes from a historical perspective how these drive the technological system forward. He does not, however, explain relations between them. In the Third National Congress for the History of Technology, in his paper, 'On the Development of the Technological System,' Huang further analyzes the process of the development of the technological system as a whole. The develoment of technology, in his view, can be divided into three stages: (a) the stage of manual means, (b) the stage of the machine, and (c) the stage of information. He then examines the system of technology at each stage.

Ke Liwen, in a paper on 'The Law of Network Reaction in Technological Development,' proposes the concept of 'elementary technology' to indicate the basic component parts needed for the realization of most substantial technological activities in the technological system, and he suggests that these resemble the cells of an organism. Switching metaphors, if the technological system can be thought of as a net, the elementary technologies constitute the mesh. The technological system thus has a hierarchical structure, as follows: technological system → technological complexes → components of a technological complex → ... → elementary technologies. The role of each type of technology in the system is not the same. Some technologies have a decisive effect on other technologies, even on the whole technological system. Such a technology is, in Ke's words, a 'leading technology,' which will be replaced by a new one when social needs conflict with the existing technological system. The emergence of a new leading technology causes a network revolution in relevant types of technology. An invention meets some social needs and thereby brings about new social needs. Thus the reaction, social needs → invention → new needs → new invention, also takes place over the network of elementary technologies, building up a new system of technology.

Seeking a set of indices to measure the development of technology is the aim of Deng Honghai's recent research. In a short but interesting essay, 'Thinking and Forecasting in the Philosophy of Technology,' Deng regards technology as 'the means and media that are used to exchange materials, energy, and information between human beings and nature.' In practical and cognitive activities, materials, energy, and information transmit the functions of all the organs of the human body to nature, making it change according to human purposes. Thus the efficiency of human practices and cognition depends in a number of different respects on the properties of technology and the ways it transmits all kinds of functions. Some relevant factors are: the power of transmission in magnifying the power or the functions of the human body; the sphere of transmission or the range and depth to which human functions are transmitted; the precision of transmission, that is, the extent to which transmitted functions of the human body correspond with human purposes; the mode of transmission in which functions are transmitted separately or together; and the use of the transmission to process, reproduce, or create materials. On the basis of these five properties, Deng divides the development of technology into three stages: (a) manual technology, (b) industrialized technology, (c) biological technology, and he then compares the characteristics of each stage.

Almost all Chinese philosophers believe that there are objective laws revealed in the development of technology. The question, in what sense or at what level technological development is in accord with subjective law, still needs to be answered. In addition, studies of the dynamic mechanisms of historical technological development are quite weak in comparison with historical laws.

Technology and Society

The relation between technology and society is quite complex. Here the problems that Chinese scholars are interested in can be indicated under two headings: (a) the relationship between the new technological revolution and social development, and (b) the relation between traditional Chinese culture and technological development. Under the first heading, problems of the impact of the new technological revolution on the division of labor, the mode of labor, productive forces, and the human sense of value, as well as ethical problems created by the technological revolution, have been hotly debated – with special attention directed to the issue of the division of labor.

In the Marxist view, the division of labor, especially the traditional division of labor which limits a worker's whole life to one simple physical profession, exercises an unjust constraint on human nature. It limits the sphere of a worker's activities, separates the worker from realizing the full benefit of his labor, and causes the inequality among workers in human society.

What is the impact of the technological revolution on the division of labor? Some Chinese scholars think that the technological revolution will itself eliminate the division of labor and return to human beings the control of their labor. Qin Qingwu (1985) argues, first, that since the new technological revolution is an information revolution based on a high stage of scientific and technological development, the boundary line between intellectual and physical labor will be eliminated; second, that the overall automation of the mechanical system and the emergence of intelligent robots will enable workers to become masters of machines instead of the slaves, organs, and appendages of machines; and, third, that the technological revolution reduces the intensity of labor, enabling workers to do whatever work they like. So, in Qin's view, the new technological revolution will eliminate the division of labor and contribute to liberating the human being from the shackles of labor. Lin Jian (1986) proposes a viewpoint similar to Qin's, but for different reasons.

Other scholars, however, argue that the technological revolution deepens the division of labor. In his essay, 'The New Technological Revolution Deepens Professional Divisions,' Hao Zhensheng opposes the view of Qin. Hao thinks that the new technological revolution is the result of a continued deepening of the division of labor and causes a further deepening. For example, under the influence of the technological revolution, the emergence of many interdisciplinary studies indicates that the division of labor is being carried further. The new technological revolution, however, changes the division of labor from the old-style division to a strictly professional one. In the ideal society (the communist society), people should be able to choose and change their occupations as they please, but this does not mean that everyone will be able to know or do everything. Without a profession, human value will not be able to be realized. The debate about the technological revolution and the division of labor has not subsided. Further inquiry into this and other problems, such as the impact of technology on relations of production, or mode of labor, can contribute to understanding the role of the technological revolution in contemporary society.

The relation between traditional Chinese culture and technological development has attracted wide scholarly attention since a symposium on 'The Causes of China's Backwardness in Science and Technology in Modern Times' was held in Chengdu in 1982. 'Traditional culture and modernization' has been an issue debated heatedly in Chinese academic circles for several years. To the question why Chinese science and technology fell behind the West in modern times, Chinese scholars have proposed various answers. Jin Guantao and others (1983) compare the Chinese technological system with that of the West. The 'open system of technology' of the West, which was gradually built up after the sixteenth century, is characterized by a technology that can be dissociated from its maker or user and easily mastered by others. Two preconditions for building up such a system of technology are economic power and a certain view of nature (Jin and others call it a 'structured view of nature'). But in China, not only was economic power minimal, but also the view of nature was organic (implicit, abstract), not structured, which precluded the growth of an open system of technology. Consequently, some Chinese technologies regularly disappeared with the death of the craftsmen making and using them.

Diao Peide and Li Xiugo (1983) have also studied the conflict between the spirit of modern technology and traditional Chinese feudal culture. For example, in Chinese feudal consciousness, people desired that 'ten thousand Guan (a kind of monetary unit) be stored at home' (a Chinese saying) – which did not encourage the investment of capital. Similarly, another saying, 'There is no need to associate with other people for doing anything,' was looked upon as the social ideal – not the division of labor and cooperation required by the modern system of technology. Similar to Diao and Li, Liu Ji (1983) has analyzed the character of the Chinese nation and its impact on the development of science and technology in China.

Other scholars have studied the impact of Chinese science and technology on Chinese culture. Hua Daming, in his 'On the Scientific and Technological Causes of the Difference between Chinese and Western Painting,' analyzes the different developments of science and technology in the East and the West, and their differential impact on painting. For example, light and shadow are mainly used in Western painting to portray the image of an object. In traditional Chinese painting, however, the main technique of expression is line drawing since an explicit optics or science of color did not exist in ancient China.

Some scholars, however, do not think that Chinese traditional culture is in any essential conflict with the spirit of modern technology. Kang Rongping (1986) argues that because of differences in the natural environment, language, or culture, each nation in the world has naturally formed its own system of technology with different national characteristics. For instance, Chinese traditional architecture is influenced by Chinese aesthetic standards. So it is not necessary to change traditional culture completely (something which is impossible anyway) just because the Chinese technological system is different from that of the West. On the contrary, when foreign technology is imported, an important thing to be considered is whether or not it can be nationalized – that is, whether it can be absorbed by the existing system of technology with its own national characteristics. To a great extent in the past

Chinese blindly copied the foreign system of technology, neglecting nationality and the nationalization of technology. Since this actually hindered the development of technology in China, Kang appeals to the Chinese to establish 'a technological system with Chinese characteristics.' Dong Shiyi, in his 'An Exploration of How to Form a Technological System with Chinese Characteristics' and other essays, outlines the main features of a technological system with Chinese characteristics and the possibility of establishing it.

Along with the two issues mentioned, many philosophers have also studied problems related to the social function of technology, and the social assessment of technology (including critiques of optimism and pessimism) – issues which must be set aside in the present overview.

Technology and Nature

In the discussion concerning technology and nature, two approaches reflect two different investigative purposes. One involves research on the natural or ecological consequences of specific technologies from a philosophical perspective; the other entails analysis of human activity from the perspective of the human technology-nature relationship, namely, studies in so-called 'general technology.'

The peculiarities of artifice, the relation between artifice and nature – sometimes called the relation between 'artificial nature' and 'natural nature' – and other 'ontological problems of artifice' have been studied by philosophers, though not yet at any great depth. Chen Changshu, in his essay, 'A Tentative Study of Artificial Nature,' considers the position and effect of artifice from the perspective of the humanity-nature relationship. He argues that 'nature which has been transformed, molded, and processed should be called "artificial nature." ' Yuan Deyu and Chen Changshu (1986) similarly distinguish artificial nature and natural nature, analyzing some peculiarities of artifice and maintaining that the operations, laws (or patterns), and methods of forming of artificial nature should be studied in the philosophy of technology. Chen Nianwen (1984) analyzes the process of the forming and the development of artificial nature.

The relation between humanity and nature varies according to the development of technology and raises important questions. What was the relation between humanity and nature in the past? What will happen to this relationship in the future? What should be the relationship? Bian Chunyuan, in his essay, 'Humanity – Technology – Nature,' argues that technology should not be in conflict with but coordinated with nature. Hua Daming and Hua

Liguang (1983) analyze the relation between humanity and nature from an evolutionary perspective. They think that the relationship has gone and is going through three main periods: (a) an organic and unified relationship between primitive humanity and nature; (b) a relation of opposition in which humanity seeks to conquer nature; (c) the re-establishment of a new organic and unified relationship.

How to coordinate humanity and nature is studied by Yuan Deyu and Chen Changshu in *On Technology*. They argue that the relation of opposition brought about by technology can and should be overcome by using technology itself. Chen Nianwen (1984) emphasizes, however, that to solve the problem of the relation between humanity and nature, a concept of nature as a whole must be formed, and a social organization which can arrange production and distribution in a planned way must be established.

Studies in so-called 'general technology' have been initiated by Liu Zeyuan. In his 'The Technological Category: The Dynamic Role of Humanity Dominating Nature' and other essays, Liu argues that the essence of technology is the dynamic role of humanity dominating nature, which exists not only in human productive activities but also in social and intellectual activities. Therefore, a new discipline is necessary – general technology – in which the dynamic rule of humanity over nature that exists in material production, social life, and intellectual development can be comprehensively revealed. The three main branches of general technology are 'the technology of material production,' 'the technology of social construction,' and 'the technology of intellectual production.'

The basic task of the technology of material production, in Liu's view, is to apply general laws discovered in natural science to direct the forces of material production by means of particular rules in technological science and technological objects created in the course of the development of technology. The basic task of the second branch – technology of social construction – is to change the general laws of society revealed in basic social sciences (such as political economy) into social engineering and technology and then further into social substance by means of social technological science. The object of study in this branch is not society alone but the 'society-humanity-technology-(special) nature complex.' The basic task of the third branch – technology of intellectual production – is to inquire into process, means, and method of intellectual production. Liu thinks that the essential factors in the process of human intellectual labor, which is similar to physical labor, are: (a) cognitive subject – the intellectual producer and his activity with certain purposes; (b) object of cognition – the objective world (nature, man, society);

(c) means of cognition – material means and methods in intellectual labor. In Liu's view, intellectual labor is a special kind of practical activity. Liu's philosophy of technology goes beyond the scope delimited by most Chinese philosophers of technology, expanding into problems of philosophy of science and social philosophy, which, Liu stresses, should be studied from the perspective of technology.

The Methodology of Technology

In regard to the methodology of technology, Chinese philosophers have tried, by studying the particular methods used in all types of specific technologies, to describe a kind of general technological method which can be used in all technological development activities. In his essay, 'Toward a Methodology of Technology,' Gao Dasheng appeals to philosophers to study the methods used in the course of the development of technology. The historical development of technological method and the difference between scientific and technological method are analyzed in his essay.

In 'A Tentative Study of the Methodology of Technology,' Zhang Xielong further distinguishes the basic forms of general technological methodologies. According to the sequence of their historical development, he divides the general methods of technology into three types: (a) basic technological methods – such as imitating, transplanting, or simulating; (b) methods by which scientific theories are materialized into technology – such as formulating, modeling, or testing; and (c) methods of systems engineering – such as the three-dimensional matrix, or design by the human-machine system.

Problems in the second type of technological method, such as those of modeling, have been studied more deeply than other types. Gao Dasheng's 'On the Method of Modeling' analyzes the philosophical presuppositions of modeling, its basic nature, and the stages of its development. In his essay, 'A Study on the Application of Conceptual Models in Engineering,' Wang Hongbo investigates conceptual modeling in engineering, its relation to scientific method, its role in engineering design, and other problems.

Other kinds of technological methods have also been studied separately. According to Wang Haishan's classification (1984), these investigations can be put into three categories: (a) research on the methods of technological planning; (b) research on the methods of technological research; and (c) research on the methods of technological design. Chinese philosophers attempting to offer effective methods to engineers and technologists think that only such investigations in the methodology of technology are significant. This, however, cannot be maintained because of the separate investigations of the different types of technological methods; technologists cannot know in which instance and on what problem each method should be used. How to unite technological methods philosophically in order to offer a useful methodology to technologists remains an important problem confronting Chinese philosophers. The solution of this problem requires further understanding of the essential character of technology, returning philosophers to the first problem: the essence and essential factors of technology.

3. CONCLUDING REMARKS

We have introduced the five important issues discussed in Chinese philosophy of technology. But Chinese philosophical research on technology is not confined to these five issues alone. The aesthetics of technology is another flourishing discussion. Chinese philosophers have been paying much attention to research in aesthetics since the 1950s. In addition, brain-machine relations and other philosophical problems in cybernetics and systems theory attract the interests of some Chinese philosophers. Indeed, there are so many studies in this area that it is difficult to introduce them in a comparatively short article. Most Chinese philosophers, however, do not think that these two issues are part of the philosophy of technology – the task of which, in the Chinese perspective, remains somewhat different from that in the West.

To some extent, any philosophy of technology is always going to be influenced by the philosophical tradition of the country in which it takes place. Since all Chinese philosophers are Marxists, dialectical and historical materialism provides a theoretical basis and guidance for their research in the philosophy of technology. In methodology, dialectical materialism stresses that a complex object should be studied not analytically but comprehensively and a historical phenomenon should be studied not statically but dynamically. Thus, most Chinese philosophers attempt to grasp technology as a whole and to study it from dynamic and evolutionary perspectives. Few Chinese philosophers would consider technology only as a material means, or as applied science, or as a kind of skill since most of them have a broad concept of technology. That they emphasize studies of the essential factors in technology and their relation with each other demonstrates their systematic perspective.

At the same time, research in philosophy of technology is also influenced by traditional Chinese philosophy in which humanity is considered as a part of nature, in which humanity and nature are regarded as a whole, and in which harmony is stressed at the expense of conflict. Therefore, the 'coordinate development of technology and nature' and the 'coordinate development of technology and society' can often be found in Chinese works on the philosophy of technology. The emphasis of traditional thought on 'applications' has also affected the goals of research in Chinese philosophy of technology. Chinese scholars desire that studies of the essence and essential factors of technology and patterns of technological development should provide a theoretical basis for the formulation of technological policy. Research in the methodology of technology should be useful for technologists in inventing, creating, designing, and other activities.

Research in the philosophy of technology is also influenced by the stage of social development. China is a developing country, so how to develop its science and technology is an important problem confronting it. The effect of this problem on research in the philosophy of technology is that great attention has been paid to the patterns or laws of the development of technology. Some even think that the philosophy of technology is only a discipline seeking the laws of technological development. Compared with Chinese philosophers, Western philosophers have paid more attention to the social and ethical consequences of technology because of the differences between developing and developed countries.

Although the problems stressed by Chinese philosophers are different from Western philosophers, both have a poor understanding of technology. What is technology? is a question faced by all countries in the world, developing and developed. The ways in which technology is studied will undoubtedly be different, depending on social and cultural backgrounds. The existence of differences is actually a benefit to developing the philosophy of technology. Meanwhile, the exchange of views between various traditions of research ought to be strengthened. In China, this has been emphasized since the inception of the philosophy of technology. Japanese research in the philosophy of technology has been known intimately by Chinese scholars. German and American research is also being introduced into Chinese academic circles. Chinese philosophers hope that through the efforts of all philosophers a comparatively explicit and comprehensive understanding of technology will develop in the future.

REFERENCES

[*Note:* Some liberties have been taken with the transliterations of titles provided by the authors; insofar as possible, they are rendered the same way here as in the text – *Editor.*]

Bian Chunyuan. 'On the Essence of Technology,' in *World-Outlook and Methodology in Natural Science* (Qiushi Press, 1983).

Bian Chunyuan. 'Humanity – Technology – Nature: On the Coordinate Development of Technology and Nature,' *Journal of the Beijing Iron and Steel Engineering Institute (Social Science)* (1986, no. 1).

Chen Bing and Xie Shusen. 'A Study of the Effects of the Scientific and Technological Revolution on Productive Forces,' *Philosophical Research* (1985, no. 2).

Chen Changshu. 'On Research in Philosophy of Technology,' *Journal of the Northeast Institute of Technology* (1983, no. 1).

Chen Changshu. 'A Tentative Study of Artificial Nature,' *Philosophical Research* (1985, no. 1).

Chen Changshu. 'Some Problems in the Philosophical Study of Technology,' in *Proceedings of the First National Congress of Philosophy of Technology* (Chengdu, 1985), Chinese Society for Dialectics of Nature (1985).

Chen Changshu. 'Technology as an Object of Study for Philosophy,' *Journal of Dialectics of Nature* (1985, vol. 7, no. 3).

Chen Changshu. 'The Impact of Technology on the Development of Philosophy,' *Studies in Dialectics of Nature* (1986, no. 6).

Chen Fan. 'On the Essential Factors of Technology and Its Structure,' in *Proceedings of the First National Congress of Philosophy of Technology* (1985).

Chen Junhong. 'Some Problems in the Study of the New Technological Revolution and the Division of Labor,' *Philosophical Research* (1986, no. 8).

Chen Nianwen. 'Humanity and Nature,' in *The Principle of Natural Dialectics*, edited by the Chinese University for Science and Technology, Hunan People's Publishing House (1984).

Chen Wenhua, and Gu Zuxue. 'On the Definition and Character of Technology,' *Information from Nature* (1985, no. 1).

Den Honghai. 'A Preliminary Inquiry into the Periodic Law of Technological Development,' *Philosophical Research* (1984, no. 12).

Den Honghai. 'Thinking and Forecasting in the Philosophy of Technology,' *Encyclopaedic Knowledge* (1986, no. 6).

Deng Shuzeng. 'The Character and the Objects of Philosophy of Technology,' *Information from Nature* (1984, no. 3).

Deng Shuzeng. 'The Scope and Theoretical Frame of Philosophy of Technology,' in *Proceedings of the First National Congress of Philosophy of Technology* (1985).

Diao Peide, and Li Ziugo. 'The Introduction of Modern Technology and Its Conflict with the Chinese Feudal Cultural Tradition,' *Studies in Science of Science* (1983, no. 20).

Dong Shiyi. 'The Main Signs of a Technological System with Chinese Characteristics,' *Reference Material on Economic Research* (1983, no. 185).

Dong Shiyi. 'An Exploration of How to Form a Technological System with Chinese Characteristics,' *Studies in Science of Science* (1983, no. 2).
Gao Dasheng. 'On the Method of Modeling,' *Philosophical Research* (1981, no. 7).
Gao Dasheng. 'The Strategy of Technological Development and Dialectical Thinking,' *Philosophical Research* (1985, no. 3).
Gao Dasheng. 'Toward a Methodology of Technology,' *Philosophical Research* (1986, no. 11).
Guan Jintang. 'On the Definition of Technology,' *Information from Nature* (1985, no. 1).
Hao Zhensheng. 'The New Technological Revolution Deepens Professional Divisions,' *Philosophical Research* (1985, no. 10).
He Zhongxiu, and Guan Xipu. 'Some Problems about Research on the Law of Scientific and Technological Development,' *Science, Economy, Society* (1983, no. 2).
Hong Xiaotao. 'On the Law of Accelerated Development of Technology,' *Information from Nature* (1985, no. 4).
Hong Xiaotao. 'Thoughts on Philosophy of Technology,' *Journal of Anhui University (Social Science)* (1986, no. 4).
Hu Xiangming. 'The Role of the Technological Revolution and Its Influence on Society,' *Philosophical Research* (1984, no. 10).
Hua Daming. 'On the Scientific and Technological Cause of the Difference between Chinese and Western Painting,' in *Scientific Tradition and Culture*, edited by the editors of the *Journal of Dialectics of Nature*, Sanxi Science and Technology Press (1983).
Hua Daming, and Hua Liguang. 'The Origin and Development of the Alienation of Science,' *World Science* (1983, no. 5).
Huang Linchu. 'Problems Studied by the Philosophy of Technology,' *Newsletter of Natural Dialectics* (1982, no. 12).
Huang Linchu. 'The Essential Factors and Structure of a Technological System,' *Information from Nature* (1984, no. 2).
Huang Linchu. 'The Forming of a Technological System and the Pattern and Trend of Its Development,' in *The Principle of Natural Dialectics*, edited by the Chinese University for Science and Technology, Hunan People's Publishing House (1984).
Huang Linchu. 'On the Development of a Technological System,' in *Proceedings of the Third National Congress of History of Technology* (Huangsham, 1984), Chinese Society for the History of Science and Technology (1984).
Ji Yuxing. 'Toward a Philosophy of Artificial Nature,' *Study and Research* (1985, no. 1).
Jin Guantao, Fan Hongye, and Liu Qingfeng. 'The Cultural Background and Evolution of a Scientific and Technological Structure,' in *Scientific Tradition and Culture*, edited by the editors of the *Journal of Dialectics of Nature*, Sanxi Science and Technology Press (1983).
Kang Rongping. 'Toward a Science of Technology,' *Natural Dialectics Review* (October 9, 1983).
Kang Rongping. 'The Law of Acceleration of Technology Transfer,' in *Proceedings of the Third National Congress of History of Technology* (1984).

Kang Rongping. 'Establishing a Technological System with Chinese Characteristics,' *Studies in Dialectics of Nature* (1986, no. 2).
Ke Liwen. 'The Law of Network Reaction in Technological Development,' *Information from Nature* (1985, no. 1).
Ke Liwen. 'Thoughts on the Relative Independence of Technological Development,' in *Proceedings of the First National Congress of Philosophy of Technology* (1985).
Kou Shiqi et al. 'On the Evaluation Criterion for Technological Theory,' *Research on Nature* (1986, vol. 5, no. 2).
Li Huiguo, and Wu Yuanliang. 'The Revolution of Contemporary Technology and the Modernization of Social Science Research,' *Philosophical Research* (1984, no. 6).
Li Pengcheng. 'The New Task Set by the New Revolution in Technology,' *Philosophical Research* (1984, no. 6).
Lin Jian. 'On the Old-Style Division of Labor and Its Passing Away,' *Philosophical Research* (1986, no. 8).
Lin Youji. 'The Relation between Man and Nature in the Technological Revolution,' *Fujian Forum (Literature, History and Philosophy)* (1984, no. 4).
Liu Dongzhen, and Yang Derong. 'On Technological Systems,' *Information from Nature* (1983, no. 3).
Liu Ji. 'One Cause of China's Backwardness in Science and Technology in Modern Times,' in *Scientific Tradition and Culture,* edited by the editors of the *Journal of Dialectics of Nature,* Sanxi Science and Technology Press (1983).
Liu Lixian, and Zhang Jia. 'Changing Classes, from Workers to Intellectuals: The Necessary Law of Historical Development,' *Philosophical Research* (1985, no. 6).
Liu Zeyuan. 'The Technological Category: The Dynamic Role of Humans Dominating Nature – On General Technology,' *Studies in Science of Science* (1983, vol. 1, no. 2).
Liu Zeyuan. 'The Essence of Technology and the Methodological Principle of Technological Development,' *Information from Nature* (1984, no. 1).
Liu Zeyuan. 'Social Development and Social Technology,' *Studies in Dialectics of Nature* (1986, vol. 2, no. 1).
Lo Kuang. 'On Technological Categories,' *Journal of Anhui University (Social Science)* (1985, no. 1).
Lu Pinyue. 'A Theory about the Systematic Evolution of Technology,' *Journal on the Potential of Science* (1983, no. 2).
Pan Liangtao. 'The Concept of Technology,' *Newsletter on Studies of Natural Dialectics* (1981, no. 1).
Pan Shuming. 'Some Problems about the Contemporary Technological Revolution and Modes of Production,' *Philosophical Research* (1985, no. 3).
Qin Qingwu. 'On the Technological Revolution and the Old-Style Division of Labor,' *Philosophical Research* (1985, no. 6).
Qin Qingwu. 'Some Problems about the Professions and the Old-Style Division of Labor,' *Philosophical Research* (1987, no. 5).
Shen Mingxian. 'The Social Role of Science and Technology: A New Antinomy,' *Philosophical Research* (1987, no. 5).
Shi Guozhu. 'On Technological Thinking,' *Scientology and the Management of Science and Technology* (1983, no. 3).

Shi Yiqing. 'The Impact on Social Development of the Appearance of Robots,' *Philosophical Research* (1984, no. 9).
Song Huichang. 'Some Ethical Problems of Contemporary Technological Development,' *Philosophical Research* (1985, no. 3).
Sun Shuping. 'Technological Revolution and Social Revolution,' *Journal of Nanjing University* (1980, no. 3).
Wang Haishan. 'The Process and Methods of Technological Invention,' *Fujian Forum (Literature, History and Philosophy)* (1983, no. 1).
Wang Haishan. 'Methods for Generating Technological Principles,' *Information from Nature* (1984, no. 1).
Wang Haishan. 'A General Method for Engineering and Technology,' in *The Principle of Natural Dialectics,* edited by the Chinese University for Science and Technology, Hunan People's Publishing House (1984).
Wang Haishan. 'The Principle of Creativity in Technological Invention,' *Information from Nature* (1986, nos. 1–2).
Wang Hongbo. 'A Study of the Application of Conceptual Models in Engineering,' *Philosophical Research* (1984, no. 12).
Xie, Enze. 'The New Technological Revolution and Social Development,' *Journal of Northeast Normal University (Social Sciences)* (1984, no. 5).
Yan Kangnian. 'Some Problems in the Study of Technological Revolutions,' *Journal of Dialectics of Nature* (1985, vol. 7, no. 3).
Yan Kangnian. 'Inquiry into Some Fundamental Problems in the Study of the Theory of Technology,' in *Proceedings of the First National Congress of Philosophy of Technology* (1985).
Yang Derong. 'Reflections on Studying the Law of Technological Development,' in *Proceedings of the Third National Congress of History of Technology* (1984).
Yuan Deyu. 'Technology as an Independent Object of Study,' *Newsletter on Natural Dialectics* (1982, no. 2).
Yuan Deyu. 'The Scope and Character of the Philosophy of Technology,' *Information from Nature* (1982, no. 2).
Yuan Deyu, and Chen Changshu. *On Technology*, Liaoning Science and Technology Press (1986).
Zhai Zhihong. 'The Difference between Science and Technology in Terms of Their Origins,' *Information from Nature* (1983, no. 4).
Zhang Letong. 'Similarities and Differences between the Methods of Technology and Science,' *Information from Nature* (1984, no. 1).
Zhang Xielong. 'A Tentative Study of the Methodology of Technology,' *Philosophical Research* (1985, no. 6).
Zheng Shiming. 'On the Historical Character and Developmental Causes of Modes of Production,' *Philosophical Research* (1985, no. 1).
Zou Shangang. 'On the Concept of Technology,' in *Proceedings of the First National Congress of Philosophy of Technology* (1985).
Zou Shangang, Peng Jinan, Su Ziyi, et al. *Some Dialectic Contents in Engineering and Technology,* The People's Educational Publishing House (1979).

WOJCIECH GASPARSKI

DESIGN METHODOLOGY: A PERSONAL STATEMENT

1. INTRODUCTION: WHAT I MEAN BY 'METHODOLOGY'

I believe that *practice* can be clarified with just as much exactness as theory; indeed, I would go further and say that the study of design methodology will significantly improve practice – and one ought to recall that doing science, even theoretical science, is a practical activity which we must do *by design*. With this in mind, I turn first to the definition of the term 'methodology.'

The term 'methodology' is often used in two diametrically opposed senses.[1] They are well characterized by Mark Blaug in an informative book on methodology of economics:

> A fatal ambiguity surrounds the expression, 'the methodology of....' The term methodology is sometimes taken to mean the technical procedures of a discipline, being simply a more impressive-sounding synonym for methods. More frequently, however, it denotes an investigation of the concepts, theories, and basic principles of reasoning of a subject, and it is with this wider sense of the term that we are concerned in this book. To avoid misunderstanding, I have added the subtitle, *How Economists Explain*, suggesting that 'the methodology of economics' is to be understood simply as philosophy of science applied to economics (Blaug, 1982, p. xi).

The case is similar with the methodology of design. One finds studies in which 'the methodology of design' refers simply to methods – sometimes *the* method – of design and the fancy phrase is used just to make the study 'more scientific.' I have no objection to such studies as long as the authors are not guilty of confusing their readers by equating methodological studies *par excellence* with the technicalities of designing – or as long as these authors do not insist on their interpretations or insist on 'practical applicability' as a criterion.

The methodology of design as I understand it parallels Blaug's methodology of economics – that is, it is philosophy of science applied either to all practical disciplines (the usage of Tadeusz Kotarbinski)[2] or to applied science (Mario Bunge)[3] or to the sciences of the artificial (Herbert A. Simon)[4]. Understood in this fashion, the methodology of design is neither the practice of designing nor instructions for that purpose but theoretical reflection on the design process along the same lines as philosophers of science reflecting on methodology. In this connection, anyone proposing to do a study of the methodology of design might well follow Blaug's example and add a subtitle: 'How Designers or Researchers in Practical Sciences Deal Logically with Change.'

Turning to the concept of 'design,' it is not a datum given *a priori*, not something demanded by the theory of design – as Myron Tribus notes:

> It is important to realize that a theory is *constructed,* not *discovered.* That is, a theory can be developed which will contain sufficient concepts to encompass a subject of interest and provide connections among these concepts. ... When properly understood, a theory defines its domain of application. Thus, thermostatics applies only to equilibrium – and equilibrium is defined only *via* thermostatics. This apparent circularity is inherent in all theories. It should not be regarded as a deficiency (Tribus, 1969, p. 382).

To complete the picture, it should be added that the methodology of design – understood as the *sui generis* theory of methodological design problems – deals with design defined on its own terms (see Gasparski, 1978, p. 45). Thus, a definition of design is not a question external to the methodology of design; its discussion should not be relegated to a separate introduction.

There is one more problem – tangential rather than pertaining to the substance of the methodology of design – that still needs discussion. It has to do with the *program* or structure of the study. The methodology of design is one of several disciplines that study design just as the methodology of the sciences is one of several disciplines – sometimes called 'metascientific' disciplines – that study the progress of human knowledge. But the phrase 'methodology of science' is used in a variety of ways depending on the particular programmatic intentions of particular methodologists or of the scientific schools of thought with which the methodologists are connected.[5] Among programs for the methodology of science, there is one which focuses on the methodology of methodology itself; this is the program of Kazimierz Adjukiewicz (1974, pp. 173–177). That program, along with the methodological reflections of Kotarbinski (see Gasparski, 1983, pp. 23ff.), provides a strong foundation for the methodology of design[6] (see Gasparski, 1978). The approach combines *praxiology* with systems-based concepts of advances in design (Gasparski, 1981, pp. 11ff.); it combines the praxiology of methodology in general with general systems theory and so-called 'science of science.'

The methodology of design includes two parts, paralleling the two parts of the methodology of science according to Adjukiewicz. The first part, 'pragmatic methodology,'[7] deals with the analysis of the purpose of design, defines its essence, and analyzes the procedures applied in the process of design. This part is also called 'methods of design' and includes a description of individual actions involved in creating a design. The second part, 'apragmatic methodology,'[8] focuses on the object of design and particularly on the language in which are formulated both problems and their solutions considered as complex objects or systems.

Depending on the extent of the lawfulness and validity of the theses of design methodology, one can distinguish general design methodology from particular design methodologies.[9] The place of design methodology within the framework of methodological knowledge in general is shown in Figure 1.

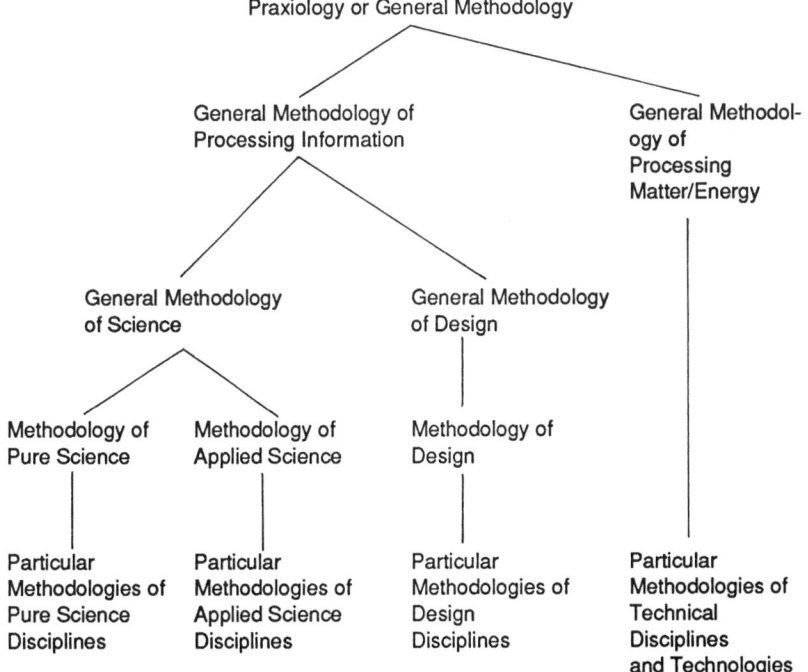

Figure 1. *Design Methodology within the Framework of Methodological Knowledge in General.*

2. DESIGN: THE OBJECT OF DESIGN METHODOLOGY DISCUSSIONS

We are now ready to enter through the gates leading to the mansion of design methodology, and once inside we find ourselves in a great hall. Behind this hall lie the living rooms – pragmatic and apragmatic – whence the lady of the house hurries to meet her guests and present herself. As well-mannered guests, we should remove the outer garments of the everyday ways in which we are accustomed to thinking about design. We need to allow the lady to introduce herself, even though – conscious of the polymorphism of her existence – she is not inclined to do so. The disinclination reminds us of Ernest Gellner's (1979, p. 51) phrase: the complexities extend 'from the actual use of words to the answer to philosophical problems or from a conflict between the actual use of words to the falsity of a philosophical theory.'[10]

Design in the broad sense includes two notions: (a) design as a method, and (b) design as an activity. The former belongs in a special way to the practical sciences. The activity of producing designs – design in the strict sense – is really a pre-activity, a preparation getting ready for other activities to be carried out in a certain determined way. The two notions have the common aim of creating practical models – that is, models of states of affairs someone wants to have realized. The two differ in terms of the minute detail of the models: the practical sciences are concerned with general models of the states of affairs, whereas the activity of creating designs involves the production of detailed practical models applicable to units or repeatable states of affairs.

In 1913 design methodology had already become an object of interest to Kotarbinski, who said it is the only methodologically justifiable use of *hypostasis:*

Questioned about the future, a practical person will invoke the criterion of his own desires, and that is the standard he will use in his reply. He will do the following: he will evoke within himself an inclination involving a lively emotion; then he will begin, led by that emotion, to evoke within his inner world a creative image of a desired object. Starting with a primitive, clumsy shape, it will end up at last being put in some desired shape. For instance, he might form the image of a statue that he wants to make, of a song that he wants to sing later on, of a design that he wants to construct, to 'realize,' later on. After developing the image, he will look at the future with the eyes of imagination, and the way he will describe the future is in terms of his vision. And when he looks at the future subject to his desires, his methods will be better than those of the typical theoretician. He will do things that theoreticians are not accustomed to doing, that indeed they are reluctant to do – namely, he will take the object of his fantasy to be the pattern of 'objective reality.' He will judge the course of future actions on the basis of features of objects seen in the world of fantasy, which he will then transfer to the world of objective space, making a hypostasis. And in doing so he will be correct: hypostasis is always an unreliable cognitive method – except with respect to one's desired future. Here it is a good, indeed the best cognitive method (Kotarbinski, 1975, pp. 172–173).

Design methodology includes three procedures: *generating,* or envisioning future states of affairs; *modeling,* or providing descriptions of these states; and *testing,* or analyzing their feasibility.

Generating designs or concepts of future states of affairs is similar to generating the concepts involved in scientific statements; it belongs to the so-called context of discovery and, strictly speaking, it cannot be the object of methodological research. However, there have been attempts to systematize the concept-generation process. The most successful has been the 'morphological' procedure of F. Zwicky, an astronomer. Another attempt is the ARIZ algorithm approach to solving invention tasks (the acronym is derived from a Russian name) proposed by G. S. Altszuller; it has been met with great admiration and profound skepticism. And finally one should include those iterative procedures that generate subsequent approximations to a state recognized as valuable in some accepted respect.

The object of methodological research *par excellence* is the context of justification, which includes both modeling and testing.

Practical models, descriptions of the states of affairs arrived at in the generation process, are formulated in the language or languages of the particular practical sciences on which they are based.[11] For example, models of management systems (the subject of the well-known book of J. Goscinski, 1982) are based on economics; models of the planning and design approach (see G. Nadler, 1981) are based on the behavioral sciences; and models of social design or social engineering (see G. A. Antoniuk, 1983, and R. Mattessich, 1978) are related to both the social sciences and applied sciences.

Testing – verifying or falsifying – practical models is of two kinds. First, their cognitive base is tested: that is, there is an examination of the presupposed theories of the applied (or, sometimes, basic) sciences with which they are associated.[12] An extreme example would be the testing that proved the impossibility of a perpetual motion machine by showing that no relationship of presupposition could exist between such a machine and the first and second laws of thermodynamics. Testing in this sense has to do with theoretical feasibility. The second kind of test is of the possibility of putting the model into practice; that is, it is a test of the real-life conditions for putting the model into use.

This triad of generating, modeling, and testing leads to the formulation of methods of attaining practical goals better than traditional methods. Better here means more effective; in this sense, design methodology is a procedure for deriving efficiency criteria. However, more needs to be said on this matter:

> A procedure for deriving empirical criteria amounts to an explanation – an explanation of the need to use just those terms. Theoretical researchers aim at answering *why* questions; practical researchers are not interested in that since, as a rule, efficiency criteria are not satisfied when one sets out to design a theory. The practitioner does not ask why something is so, but what to do to make it so according to the criteria of efficiency – that is, what the conditions are for realizing certain values to the highest degree. He wants to design *means* that will permit realization according to presupposed efficiency criteria. In cases in which (as happens less and less often) the best ways of meeting the criteria are arrived at by trial-and-error methods, this still serves as an explanation. It makes clear, we might say, that an inventor has found intuitively a proper solution – and he can, accordingly, go on designing future undertakings of the same sort in the same way (Nowak, 1974, p. 218).

Design in the strict sense – or, as we might say paraphrasing Adjukiewicz on scientists, the designer's *craft* in playing this role – is a particular kind of activity. Among other activities, it is singled out, first of all, as *not* being auto-telic, as having its goal as part of the design activity itself. Following G. Hostelet (1947, p. 79), we can say that designing is a practical activity the goal of which is clearly defined and the results of which can be objectively tested.

Design aims at achieving a result and has informational features associated with it. With respect to the latter, design is preliminary, a pre-activity – or, in praxiological terms, preparatory activity. According to praxiological theory, an action is preparatory when it precedes other activities with which it is associated, making their performance easier – or, in some cases, possible at all. Making easier or making possible associated activities is linked to the desired results, that is, to designs in a second sense.

The content of a design includes concepts of change, of the actions needed for realization. The conceptual preparation for associated activities[13] is precisely what I mean by design in the strict sense. Anyone who is the author of an outcome in this strict sense of the term design I would call a designer. But designers are not only people for whom that is a professional occupation with 'designer' as their job title; *any* practitioner – a manager, an organizer, an educator, even a physician – is a designer if he or she plays this role.

3. THE OBJECT OF DESIGN: PROBLEMS OF APRAGMATIC DESIGN METHODOLOGY

Concepts of changes are the essence of designs, and changes are fragments of reality that nonetheless remain the object of design. To answer the question as to what these fragments of reality are, we can analyze an example. A manufacturer (M_1) produces goods (G_1) which are regularly in high demand. On the other hand, demand for the goods (G_2) produced by another manufacturer (M_2) is decreasing. To counteract his drop in income, the second manufacturer starts producing a new kind of goods (G_3) similar to those of the first manufacturer, hoping thereby to win over some of the latter's customers.

Now the situation faced by the first manufacturer is as follows. He has been producing goods that are in high demand and has therefore been producing high income. Since this is a satisfactory state of affairs, he wants it to continue. But this requires appropriate changes corresponding to the external threats to the satisfactory *status quo:* the need is to influence the behavior of customers as the new products (G_3) come onto the market.

On the other hand, the situation that faces the second manufacturer is unsatisfactory: since demand for his products is decreasing, his income keeps going down. Trying to change his undesirable situation, the second manufacturer begins to turn out the new products (G_3).

These practical situations are viewed by their subjects, the two manufacturers, as sets of facts, estimations, and assessments of facts as a result of the estimations. Such practical estimations are of two sorts, preventative and therapeutic. In the first case, an estimation of the facts provides a positive result; in the second, it is negative. Practical states of affairs, of either kind, also include 'the rest of the world,' their context. This context may be investigated as itself a practical situation different from the one it contextualizes. In our example, the situation of the second manufacturer is part of the context of the first – and *vice versa*. In every practical situation, we have a pair: a so-called 'internal' or 'core' situation, and a complementary external situation. A pair so defined, relative to any practical situation, is one of the elements that are the object of design. The pair is elemental because reality is complex; reality is a collection of practical situations with numerous objects, each of which is a core, with all the rest complementary. For instance, in our example, one must also take into account the practical situations of the various customers, of vendors, and of other manufacturers. For this reason, selecting out of the complex reality in a given practical situation which is the appropriate pair is a task as important as it is complicated. In any evaluation, we are faced with a multiplicity of perspectives as we approach an assessment of the facts (see Linstone, 1984). Regard for or conformity with this rule, unfortunately, is not a strong point among designers.

Considering the processes that generate changes over a long span of time, one sees designers aim at authenticity, at rationality, at usefulness, even at ethics or aesthetics. The expressed goal here comes under the principle of relevance (or suitability of the changes envisioned).

Though pairing practical situations, selecting them from the complex of practical situations, is the object of design, designing itself, as an intellectual activity, is not concerned precisely with the reality of the pair but with their mapping in *language* – with the design as an intellectual problem. The elemental pairs of practical situations, so mapped in the language of design (and, we should recall, that means in the languages of the practical sciences), constitute respectively the core and the complement – or the core group and the complement group – in a design problem.

A design problem is either an interrogative or a statement which can be translated or reduced to interrogative form.

The *datum quaestionis* of these interrogatives may include information on facts or assessments, but the *unknown* is the manner in which the elemental pair of the practical situation is to be formulated. Practical situations and the solutions proposed for design problems are, from a methodological point of view, *systems*.

Here it is worthwhile quoting a concise characterization of a system as proposed by Mario Bunge (1979, pp. 4–5):

Whatever its kingdom – conceptual or concrete – a system may be said to have a definite composition, a definite environment, and a definite structure. The composition of a system is the set of its components; the environment, the set of items with which it is connected; and the structure, the relations among its components as well as among these and the environment. ...

One way of characterizing the general concept of a system [though not a proper definition] is this: Let T be a nonempty set. Then the ordered triple $s = <C, E, S>$ is (or represents) a *system over* T iff C and E are mutually disjoint subsets of T (i.e., $CE = \emptyset$), and S is a nonempty set of relations on the union of C and E. The system is conceptual if T is a set of conceptual items, and concrete (or material) if T is a set of concrete entities.

According to Bunge, systems are defined at the atomic level, and this is different for different system classes. For example, the elements at the atomic level in social systems are people, not cells; the latter are elements at the atomic level in biological systems.

System S_A, defined at the atomic level A, is a well ordered triple of A-composition (C_A), A-environment (E_A), and A-structure (S_A), at time t:

$$S_A, t = (C_A, E_A, S_A), t.$$

The practical-situation pairs or sets of pairs and their implemented designs are concrete or material systems. The atomic level defining each of them depends upon the language of the particular practical science in which the design process is carried out.

4. THE DESIGN PROCESS: THE ELEMENTS OF PRAGMATIC DESIGN METHODOLOGY

The design process consists in the formulating of a design problem and its solution. The activities connected with the formulating of a problem include the identification and interpretation of the pairs, core and complement, involved in practical situations. The activities connected with the solution of a problem include decomposing or breaking the problem into parts, doing whatever is required for each separate part, and then integrating the partial solutions. I want now to give a brief characterization of each of these activities.

Formulating a design problem in such a way as to permit an adequate solution requires a mapping of the essence of the core and complement pairs of practical situations. At the same time, one expects the proper formulation of a design problem to permit solvability. The procedure of systems-identification of practical situation pairs — of separating core from complement in a systematic way — satisfies the first requirement. Interpreting the identified practical situation pairs in terms of the appropriate applied sciences fulfills the second requirement.

The identification procedure is made up of sub-procedures: measuring characteristic variable or variables; formulating a hypothesis concerning the essence, in systems terms, of the practical situation pair; testing the hypothesis; correcting the hypothesis (in most cases) and testing the corrected hypothesis; and finding the right words to describe the practical situation pair.

Describing the practical situation pair may be tantamount to a design problem, but this is not necessarily so; it may also depend on the interpretation of this description. For this reason, it is assumed that the results of the identification procedure are not necessarily produced in the language of the designer; they may be in the language of an interpreter – who may but need not be a designer.

The interpretation procedure also involves a set of sub-procedures: accepting the results of an identification procedure as a preliminary formulation of a design problem; comparing this formulation with known formulations of other design problems and assessing its solvability; (in most cases) correcting the preliminary problem formulation and analyzing the corrected version; and formulating the design problem in the designer's language.

As noted, in both procedures there is the possibility of having to correct and reassess the problem formulation. Correction and reassessment may happen as often as the interpreter or designer thinks it is necessary. However, the pragmatics of design methodology has come up with a rule of thumb: correcting and reassessing should go on as long as a formulation with subjective probability[14] does not exceed the limit probability, *and* the subjective probability of finding a solution with a higher probability is small within the available time frame and at a reasonable cost.[15]

The methodology of design does not deal with the entire routine of problem solving, filled as it is with side-tracks, going back to start all over, jumping forward and backward, and so on – filled, that is, with all the pleasant disorder that accompanies all creative activity. Design methodology as methodology in the sense explained at the beginning (i.e., as a branch of the philosophy of science) is interested in orderly procedures – that is, in sequences in which whatever happens in one step (i) is necessarily preceded by another step ($i-1$) and is the necessary condition for whatever is to follow in the next step ($i+1$). In other words, providing a report (in pragmatic terms) of the inferences involved in a design problem solution in the context of justification is tantamount to a description of the inferential structure of designing.

The form of the inferential structure of design is shown in Figure 2 in terms of a network of problem, tasks, and solutions. At the left is the design problem, and at the right the solution. In the middle are the solutions of the various subproblems into which the original problem was broken down. These aspects are designated as *tasks* and constitute the system. These tasks are to be rigorously distinguished as parts of the core – i.e., clearly separated from questions belonging to the non-core complement – and the questions posed in these tasks must be clearly decidable with a yes or no answer. The numbering of the set of tasks into which the design problem is broken down is called the *problem order*. Other dimensions of the *p/t/s* network include the *degree of disaggregation* (the maximum number of steps from the original problem to t_r) and the *degree of aggregation* (the maximum number of steps to the solution s_{t_r} – and thus to the solution of the overall design problem).

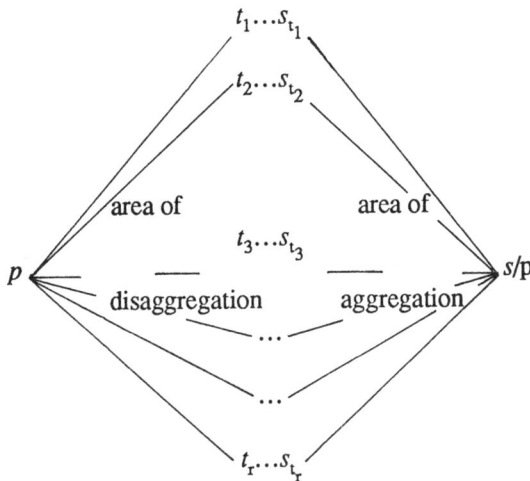

Figure 2. *The Inferential Structure of a Problem Solving Process in the Area of Design.*

I distinguish three types of networks, those with a *complete* structure, those with a *continuous* structure, and those with a *discontinuous* structure. Problems of the first sort are called *well-structured;* the other two sorts are weakly or badly structured.

During the solving of design problems, one must abide by the principle of the unity of divided design; this makes it possible to synthesize the solutions of partial problems or subproblems into a coherent and uniform solution in the systems sense. A design solution in this sense takes empirical statements from the appropriate practical sciences, recasts them, and makes them substantive in the elementary or reproducible terms needed for practical situations. The theories of applied sciences, although empirical, remain mere hypotheses (except in occasional trivial cases). Designs are substantiations of these hypotheses in the strict sense: that is, they are never the absolutely best way to realize the intention in question. The hypotheses of practical sciences are not, ultimately, the best explanations of the phenomena in question. Both the applied science hypotheses and their design substantiations are mere satisfactory solutions within the limits of the knowledge base at a given time

– including its expansion and use. In light of this, any claim to have found the 'only true solution' to a design problem must be invalid.

5. CONCLUSION: WHAT MORE DO WE NEED TO KNOW?

Design in the strict sense as defined here is not the only kind of preparation for activity. There are also investigational activities – acquiring knowledge to be used in connection with particular activities, to organize, to make decisions, etc. Common to all of these is the feature of solving practical problems. A methodological analysis of such activities using the general methodology of praxiology might lead to the formulation of a new *sui generis* discipline I would call *preparatorics*. It would be the praxiology of all preparatory activities, and its formulation might well be helped along by experience with design methodology. Certainly the latter seems to be the best impetus toward the development of this new and, I would say, hoped-for discipline (see Gasparski, 1983).

Polish Academy of Science

NOTES

[1] Kotarbinski notes that, contrary to what one might expect, lexicons of classical Greek do not include the term, 'methodology.' However, as he adds, 'Every methodological concept in current use is an extension of some understanding of "method" in classical Greek' (1961, p. 516).

[2] '[It would] perhaps be best simply to talk about critical, practical, and normative 'disciplines' (from Latin *disciplina,* from *disco,* learn), understanding a discipline to mean whatever can be taught and learned. Discussing practical disciplines, one should recall that they are exercised in conjunction with the performance of particular acts undertaken to realize a design. It could be said, for example, of a person that he or she possesses a given practical skill or art if what it is one wants to say is that the person can not only design within a given domain (that is, within a practical discipline or practical science) but can also bring the design to realization (which is what technical efficiency means) (Kotarbinski, 1961, pp. 449–450).

[3] 'Methods are means devised to attain certain ends. To what ends are the scientific method and the various techniques of science employed? Primarily, to increase our knowledge [this is the intrinsic or cognitive goal]; derivatively, to increase our welfare and power [extrinsic or utilitarian goals]. If a purely cognitive aim is pursued, pure science is obtained. Applied science [or technology] employs the same general method of pure science and several of its special methods, only applied to ends that are ultimately practical' (Bunge, 1967, pp. 25–26).

[4] 'Finally, I thought I began to see in the problem of artificiality an explanation of

the difficulty that has been experienced in filling engineering and other professions with empirical and theoretical substance distinct from the substance of their supporting sciences. Engineering, medicine, business, architecture, and painting are concerned not with the necessary but with the contingent – not with how things are but with how they might be – in short, with design. The possibility of creating a science or sciences of design is exactly as great as the possibility of creating any science of the artificial. The two possibilities stand or fall together' (Simon, 1969, p. xi).

5 'Specialists cannot agree on the main research problems of methodology and on the kind of research techniques – and especially on the nature of the conceptual apparatus – most apt for this discipline' (Wojcicki, 1982, p. 5). Though Wojcicki is talking about the methodology of the empirical sciences (where, he says, methodology 'is still not a well formulated scientific discipline'), his words provide relevant commentary on the situation with respect to the methodology of other sciences as well.

6 The place of this concept within the systematic framework of the knowledge of design science has been pointed out by, among others, Finkelstein (1982), Gregory (1980), Nadler (1980), Polowinkin (1978), Vlcek and Tondl (1983), and Archer (1981). Archer mentions design praxiology as a subfield of design research (as design science is referred to in English).

7 Adjukiewicz (1974) introduced this term (and its twin, to be mentioned in the next note) on the basis of the Greek word for 'deed.' Pragmatic methodology is the part of methodology in general that is devoted to cognitive activities.

8 Apragmatic methodology is devoted to the study of all those things that pragmatic activity aims at, including both the objects of research and the effects of research – namely, theories. (Again see Adjukiewicz, 1974, as indicated in the previous note.)

9 'Detailed design methodologies are directly linked with individual practical disciplines, also called applied disciplines. These disciplines accumulate factual knowledge about phenomena in various regions of reality that are the research object of each of them. This detailed factual knowledge, together with methodological knowledge from detailed design methodology, is used to formulate design theories' (Gasparski, 1984, p. 31).

10 This is the way people behave, as Gellner says, who have this as their strategy: 'If you cannot beat them, disqualify them! If you cannot prove rival views to be false, then say that they are meaningless! This is validation of one view by means of the exclusion of possible rivals from eligibility as candidates, in virtue of their claims having "no meaning" ' (Gellner, 1979, p. 3). The quote is from an attack on linguistic analysis only recently available to Polish readers. Gellner goes on: 'Linguistic philosophy is the buttressing of common sense by an argument based on a theory of meaning, namely that 'the meaning of an expression is its use.' It is the refusal to grant what one could call philosophic licence: the refusal to conduct philosophic discussions in a different tone, with different rules, from those ordinary discussions. It refuses to leave common sense with hat and umbrella at the door when entering into a philosophical debate. On the contrary, it makes a cult of it' (p. 54).

11 The term is used interchangeably with 'applied sciences.'

12 The relation of presupposition (or assumption) combines two scientific statements in such a way that the condition of the truth of the consequent is the acceptance of the

antecedent as scientifically true. See Bunge (1967).

[13] See W. Gasparski, 'Tadeusz Kotarbinski's Methodology of Practical Sciences and Its Influence,' in P. Durbin, ed., *Research in Philosophy & Technology*, vol. 6 (Greenwich, Conn.: JAI Press, 1983), pp. 93–106.

[14] 'The degree of conviction – i.e., the certainty of real people – I will call subjective probability. In the process of anticipation, a decision maker determines the subjective probability of hypotheses $h_1, h_2 \ldots h_m$ belonging to set H' (Kozielecki, 1975, p. 123).

[15] Costs here are understood in the praxiological sense.

REFERENCES

Adjukiewicz, K. *Pragmatic Logic* (Dordrecht: Reidel, 1974).

Antoniuk, G. A. 'Methodological Problems of Social Systems Design' (in Polish), in W. Gasparski and D. Miller, eds. *Projektowanie i Systemy*, vol. 5 (Wroclaw: Ossolineum, 1983).

Archer, B. 'A View of the Nature of Design Research,' in R. Jacques and J. Powell, eds., *Design, Science, Method* (Guildford, England: Westbury House, 1981).

Blaug, M. *The Methodology of Economics: Or How Economists Explain* (Cambridge: Cambridge University Press, 1982).

Bunge, M. *Treatise on Basic Philosophy*, vol. 4: *Ontology II: A World of Systems* (Dordrecht: Reidel, 1979).

Finkelstein, L., and A. C. W. Finkelstein. 'Review of Design Methodology,' *IEEE Proceedings* 130 (1983).

Gasparski, W. *Projektowanie: Koncepcyjne Przygotowanie Dzialan* [Design: A Conceptual Preparation for Action] (Warsaw: PWN, 1978). A partial translation is available in W. Gasparski and T. Pszczolowski, *Praxiological Studies* (Dordrecht: Reidel, 1983).

Gasparski, W. 'Studia Projekteznaweze,' in W. Gasparski and D. Miller, eds., *Projektowanie i Systemy*, vol. 3 (Wroclaw: Ossolineum, 1981). An earlier English version is available; see Gasparski, 'Praxiological-Systemic Approach to Design Studies,' *Design Studies* 1 (October, 1979).

Gasparski, W. 'The Art of Practical Problem Solving as a Subject of Scientific Exploration: An Appeal for Modern Praxiology,' in J. Calhoun, ed., *Environment and Population: Problems of Adaptation* (New York: Praeger, 1983).

Gasparski, W. 'On General and Detailed Design Research' (in Polish), *Prace Naukowe* 190 (1984).

Gasparski, W. *Understanding Design: The Praxiological-Systemic Perspective* (Seaside, Calif.: Intersystems, 1984).

Gellner, E. *Words and Things: An Examination of, and an Attack on, Linguistic Philosophy* (London: Routledge and Kegan Paul, 1979).

Goscinski, J. *Sterowanie i Planowanie: Ujecie Systemewe* [Control and Planning: A Systems Approach] (Warsaw: PWE, 1982).

Gregory, A. A. 'Deriving a Context,' *Design Studies* 1 (1980).

Hostelet, G. 'Methodology of Scientific Investigations of Human Actions' (in Polish), *Mysl Wspolczesna* 7–8 (1947).

DESIGN METHODOLOGY: A PERSONAL STATEMENT

Kotarbinski, T. 'A Theoretician and a Practitioner Approaches Future Analysis' (in Polish), in his *Selected Works,* vol. 1 (Warsaw: PWN, 1957).
Kotarbinski, T. *Praxiology: An Introduction to the Science of Efficient Action* (Oxford: Pergamon, 1965). Polish edition, 1969.
Kotarbinski, T. *Gnosiology: The Scientific Approach to the Theory of Knowledge* (Oxford: Pergamon, 1966). Based on the Polish edition, 1961; Polish original, 1929.
Kozielecki, J. *Psychological Decision Theory* (Dordrecht: Reidel, 1981).
Linstone, H. A. *The Multiple Perspective Concept* (Portland, Ore.: Portland State University, 1981).
Mattessich, R. *Instrumental Reasoning and Systems Methodology: An Epistemology of the Applied and Social Sciences* (Dordrecht: Reidel, 1978).
Nadler, G. *The Planning and Design Approach* (New York: Wiley, 1981).
Nowak, L. *Wstep do Idealizacyjnej Teorii Nauki* [An Introduction to the Idealization Theory of Science] (Warsaw: PWN, 1974).
Polowinkin, A. E. 'An Introduction by Translation Editor,' in W. Gasparski, *Praxiological Analysis of Design* (in Russian) (Moscow: Mir, 1978).
Simon, H. A. *The Sciences of the Artificial* (2d ed.; Cambridge, Mass.: MIT Press, 1981; original, 1969).
Tribus, M. *Rational Descriptions, Decisions, and Design* (New York: Pergamon, 1969).
Vlcek, J., and L. Tondl. 'Design Research in Czechoslovakia,' in W. Gasparski and D. Miller, eds., *Projektowanie i Systemy,* vol. 5 (Wroclaw: Ossolineum, 1983).
Wojcicki, R. *Wyklady z Metodologii Nauk* [Lectures on the Methodology of Science] (Warsaw: PWN, 1982).

JANET FARRELL SMITH

RESPONSIBILITY AND FUTURE GENERATIONS: A CONSTRUCTIVIST MODEL

INTRODUCTION

Questions of reproduction and biotechnology raise problems about future generations. That we do have positive responsibilities toward future generations, toward the existing next generation and toward yet to exist future generations, is one of the major premises of my discussion. Arguments can be given for this position. But my discussion here focuses on the important question of how this responsibility can be put into practice, not simply why we have it. My basic point is that, on a constructivist account, we have a basis for such responsibility in the notion of justice implicit in constitutional democracy.

My focus is on the internal structure of a theory of responsibility adequate to address the interests of future generations particularly concerning issues of equality and freedom. Such a theory of responsibility is worth addressing in its own right. To be sure, there are formidable logical-ontological problems in the status of possible, nonexistent persons, and in the alleged rights of possible persons. But even if these problems did not exist, there is sufficient reason for appeal to a principle of responsibility, rather than simply rights, for acting in the interests of future generations. One reason, for example, is that an ethic of responsibility demands a positive or active response in contrast to an ethic of rights which may merely call for noninterference.[1]

My general thesis in this paper is that a participatory democratic process involving all members of society should be the primary vehicle for social policy-making regarding future generations. In particular I argue that a theory of responsibility based on a 'constructivist model of morality' is best equipped to handle the difficult questions concerning future generations, particularly regarding biotechnologies.[2]

Both Rawls and Dworkin emphasize that the formation of a concept of justice for contemporary constitutional democracies is just that, a formation, or more precisely a construction. Justice is not a given concept inherent in natural law or intuition. It is not an *a priori* concept which is assumed antecedent to the process of construction itself. The notion of construction for both theorists involves a notion of free and equal persons as the parties involved in forming the conception of justice.

Although there are fundamental similarities, the process of construction is framed somewhat differently by each theorist. Dworkin's 'constructive model' emphasizes reflective equilibrium as the factor which allows for historical and cultural change, in contrast to a 'natural model' which draws on natural law.[3] Reflective equilibrium is taken as the collective process of a community which together settles the conception of justice. So also for Rawls. But Rawls's Kantian constructivism features his Kantian concept of the moral person, involving rational autonomy and moral objectivity. However, on each conception, constructivism is opposed to intuitionism and an *a priori* concept of natural law.[4]

Each theorist emphasizes equality. Dworkin calls it the central notion of liberalism while Rawls calls it central to the notion of a constitutional democracy. Rawls has stressed the participation of persons both free and equal in his conception of the person presupposed in Kantian constructivism. To these persons devolves the task of constructing the conception of justice.

I. SKETCH OF A THEORY OF RESPONSIBILITY

Rawls and Dworkin give what I think is an adequate philosophical basis within the present democratic arena for generating a theory of responsibility toward future generations. This basis is not yet a full theory, but contains the seeds of a fruitful development. It seems to me that we can extract some features of this constructivist model for a theory of responsibility which bypasses the thorny logical problems of the rights of nonexistent persons. These features I sketch briefly here before turning to their application to posterity in Section II, and then to issues of equality in Section III.

In his constructivist model of morality in *Taking Rights Seriously*, Dworkin has brought to the forefront a theory of responsibility which is already implicit in Rawls's *Theory of Justice*. This theory of responsibility and its particular features deserve more explicit attention. Dworkin deems such a constructivist model, in contrast to a 'natural model,' to be the one best suited to explicating the 'deep theory' behind Rawls's theory of justice and technique of reflective equilibrium. I will first summarize Dworkin's view, then show how Rawls himself elaborates a constructivist model.

The doctrine of responsibility is the crucial activator, or 'engine' as Dworkin calls it, of the selection of the principles of justice which ground the policies of the society. The reason for the central importance of this doctrine is that it is the responsibility of the members of a society to shape the ideals and policy which govern them. Three aspects of Dworkin's constructivist model are especially relevant for issues of future generations and technologies which affect their interests.

First, the model is dynamic and open-ended. Dworkin emphasizes that ideals are constructed in a social process, rather than being fixed, ahistorical essences or 'timeless features of some independent moral reality' (JR, 160). Second, the model explicitly allows that social and cultural change may affect the selection of the principles of justice. These principles are socially created, not merely discovered as one would discover the laws of physics.

Moral reasoning ... is a process of reconstructing the fundamental principles by assembling concrete judgments in the right order, mental principles by assembling concrete judgments in the right order, as a natural historian reconstructs the shape of the whole animal from the fragments of its bones that he has found (JR, 160).

Third, the constructivist model is well suited to group consideration of problems of justice, that is, to developing a theory of community rather than of particular individuals (JR, 163).

The first two features are crucial for an adequate philosophical consideration of genetic questions and biotechnologies in reproduction. Consider for example the changes in cultural views toward mentally and physically damaged offspring. In the 1923 Supreme Court Decision, *Buck v. Bell*, Oliver Wendell Holmes pronounced that, 'Three generations of imbeciles is enough.' Sterilization of such persons followed. In recent years the terminology of 'mentally retarded' has been replaced by 'mentally handicapped,' which is now being superseded by 'developmentally disadvantaged,' 'developmentally delayed,' or even 'physically or mentally challenged.' Increasing social recognition of educational and civic rights for such persons parallels the terminological change.[5] Involuntary sterilization is now prohibited and social concern focuses on rights of handicapped persons.

The general motivation behind a constructivist model is to provide a basis for generating the principles of justice. It does so by the technique of reflective equilibrium which seeks a balance between our 'ordinary, unreflective moral beliefs and some theoretical structure which might unify and justify these ordinary beliefs' (JR, 155). The main goal of the constructivist model as Dworkin sees it is to explain these ordinary moral beliefs by showing the underlying assumptions they reflect. The model of morality generated must 'provide guidance in those cases about which we have either no convictions or weak or contradictory convictions' (JR, 155), the latter being most relevant to the evolution of values in connection with technology.

I shall now turn to an analysis of four requirements which I reconstruct out of Dworkin's constructivist model of morality. These, I think, feature the central core of its theory of responsibility. Each requirement also finds a basis in Rawls's theory of justice. The requirements will not be found stated in their texts as I have named them.

(1) The Integration Requirement emphasizes responsibility for integrating intuitions and moral beliefs with the principles of justice. This requirement highlights the open-ended and flexible character of the model. Such flexibility allows for social and cultural change which would modify traditional assumptions and introduce new problems. The integration must be coherent: 'Men and women have a responsibility to fit the particular judgments on which they act into a coherent program of action' (JR, 160).

(2) The Consistency Requirement, 'a sense of responsibility for consistency in what has gone before,' flows from the assumption that it is unfair for officials to act except on the basis of a general public theory that will constrain them to consistency with the past. Consistency is understood as both internal coherence and harmony with previously established precedents. It is worth noting that the principles themselves are open to revision within the forum which enacts them.

(3) The Commitment or Sincerity Requirement prohibits the intention to deceive in making public statements. This requirement emphasizes the attitudinal aspect of the phrase 'consistency with conviction.' It requires that 'each of the men and women who reason within the model will hold sincerely the convictions they bring to it' (JR, 162). This requirement could be compared with Habermas's sincerity condition in his pragmatics of communication. Sincerity, according to Habermas, is, along with truth, one of the validity claims raised by any speech act.[6]

(4) The Accountability Requirement is the most complex and most relevant to our purposes. It involves the following four components:

(a) Moral and legal decisions must be accounted for by a background theory of justice.

(b) Decisions must be presented from a public standpoint which can evaluate differing intuitions, not from the 'personal standpoint of the individual who holds them.'

(c) Decisions must be debated publicly. 'A public standard will provide a forum for testing or debating or predicting what officials do, and not allow appeals to unique intuitions that might mask prejudice or self-interest' (JR, 156).

(d) Decisions must be open to and accommodate the convictions a community holds in common. Dworkin notes that the model is a group model of consideration and issues in a theory of community rather than one of particular individuals.

Each of the above requirements finds a basis in Rawls's discussion in *Theory of Justice* and more recent articles. On (1), the integration requirement, reflective equilibrium is described as a process of 'going back and forth' between contractual conditions, principles, and jugdments about justice. The process of mutual adjustment of intuitions about justice (e.g., that slavery is unjust), principles, and judgments occurs in hearing out conceptions of justice other than one's own initial intuitions (TJ, 20). This process can be seen as the content corresponding to the 'virtues of form' stated in requirements (3), sincerity, and (4), accountability.

The consistency requirement (2), flows from Rawls's notion of rationality and reasonableness founded on his Kantian notion of objectivity. Rawls's discussion of sincerity and commitment as examples of 'virtues of integrity' backs up the sincerity or commitment requirement in (3). These, he says, are 'among the excellences of free persons.' They lead us to make our beliefs our own and 'not adopt them as handed to us by others' (TJ, 519). The sincerity requirement includes truthfulness as well as authentic self-determination of values, which is also a kind of enactment of truthfulness in regard to values.

The accountability requirement involves responsibility for the chosen conception of justice as expressed in requirement (4a). Rawls stresses that 'in the original position the parties agree to be held responsible for the conception of justice that is chosen' (TJ, 519). The same principle holds for practices as well as theories. Those who specifically take certain actions and those in authority 'are accountable for the policies they pursue.' 'We cannot shift responsibility for what we do onto others' (TJ, 519).

Accountability involves publicity on a fundamental level as a 'natural condition' of society conceived as a social union of social unions (TJ, 582). Members of a well-ordered society 'follow and know of one another,' and 'they follow the same regulative conception' (TJ, 582). In his more recent Dewey Lectures, Rawls places particular emphasis on publicity: 'Everyone accepts and knows that the others likewise accept the same principles, and this knowledge in turn is publicly recognized' (KC, 537). This point supports the public standpoint stated in clause (4b).

To put the ideals of publicity and accountability into practice, there must be further tests of public assessment, including scrutiny of alternative conceptions of justice. Rawls calls this 'the complete justification of the public conception of justice' (KC, 537). Public debate and testing of a conception of justice as well as its practical implications relate to the theoretical level as in (4a), and the policy level as in (4c).

Let us illustrate briefly the manner in which this constructivist model might work. Consider the technologies of prenatal diagnosis applied to pregnant women: amniocentesis, sonograms, genetic screening and testing.[7] The very existence of such technologies, as biologist Ruth Hubbard has noted, exerts a persuasive and sometimes a coercive influence to utilize them.

On a constructivist model it would be a matter of public debate how and in what cases these technologies should be applied and made available on a wide scale. For in some cases and according to some theories, they further the well-being of both the woman and the potential child. For others, the presence of technology exerts a coercive influence to use it even if it is not needed. The accountability requirement demands that the various interests these technologies serve be exposed and accounted for in public debate. There would be full opportunity to challenge the necessity, for example, of repeated ultrasound diagnosis whose effects on the fetus are still unknown.

The commitment or sincerity requirement demands that the motive and goals of each party or group be stated in a way that is transparent to the public at large, that does not mask, for example, the intention of medical technology to test the efficacy and safety of the devices on the very populations which allegedly are to benefit from their use. Such examples are pursued again in Section III.

II. GROUNDING RESPONSIBILITY TO FUTURE GENERATIONS IN A FUTURE EXTENSION PRINCIPLE

In this section I suggest, briefly and without extended argument, a philosophical basis in a constructivist model for grounding responsibility for future generations. This suggestion avoids many of the difficulties on the status of possible or nonexistent persons noted in the philosophical literature by Brian Barry, Derek Parfit, Gregory Kavka, and others.[8]

To ground such a principle, we must focus on the present social arena, and upon the open-ended and dynamic character of public debate incorporated in the accountability requirement stated above. Advocacy of the interests of future generations can, without ontological difficulties, be made within this forum. Moreover, by appeals to Rawls's conception of a well-ordered society, a principle of responsibility to do so can be generated out of the constructivist model.[9]

The crucial premise is the open-ended character of the public forum persisting into the future. Rawls gives this open-ended character greater emphasis in his Dewey Lectures where he lays down the model-conception of a well-ordered society, one of the three main model-conceptions of justice as fairness. Rawls's basic claim is that there is no final date at which society is envisioned to 'wind up' its affairs. The members of such a society

> view their common polity as extending backward and forward in time over generations, and they strive to reproduce themselves, and their cultural and social life in perpetuity, practically speaking (KC, 536).

Rawls also says that a closing date for such a society is 'inadmissible and foreign to their conception of their association.'

We have, in sum, three premises. (1) No final date of society's affairs is admissible on the members' conception of their society, i.e., on the model-conception of a well-ordered society; (2) the common polity extends itself backward and forward in time by its own conception of itself; (3) the members of society strive to reproduce themselves and their cultural-social life in perpetuity.

Now, from these premises we can deduce – or, more weakly, formulate – a principle extending the social forum into the future. Reasoning by a 'future-extension principle' we could argue that the society, if it is to continue this forum of democratic participation, must provide for the interests of its present offspring, the next immediate generation, so as to be capable of participating in the forum. Then, by a kind of social principle of induction, we can conclude that provision for succeeding generations must also be included in the present consideration.

From the future-extension principle it follows that a cut-off point in some portion of the foreseeable future is arbitrary, unreasonable, and hence unjustified. Therefore, as persons are responsible now to the social polity, so they ought to be responsible for the continuing persistence of the polity into the future.

Furthermore, to buttress the argument, we could point out that the notion of a discrete generation, as Brian Barry has observed, is an abstraction from a continuous process of population replacement.[10] Any attempt to demarcate a cut-off point of consideration which relies on a number of N generations where N are discrete units – e.g., we should consider the interests of the next five generations – is as arbitrary and ill founded as setting some random date in the future. In other words, to cut off the potential for continuation of the forum is to restrict in an arbitrary and unreasonable fashion both the membership and quality of participation in future debate. A democratic social process, by its own definition, is open-ended toward the future in a way a temporary congress, for example, is not.

Restriction of the forum to considerations of provisions only for present membership would undermine the continuation of the forum itself. The process of construction involves sustaining the forum into the future as well as adjusting its principles to past agreements.

The future extension principle thus makes it logically plausible and socially necessary to address the interests of future generations, e.g., of contamination from toxic waste, nuclear accidents, and medical-biological research affecting future generations.

The difference between a constructivist model and a 'motivational assumption' within Rawls's theory should be stressed here. My argument derives responsibility from the present democratic forum interpreted as a constructivist process. It does not deduce a care and concern for the next generation from a 'motivational assumption' embedded within an original position (TJ, 128). The parties to the original position are taken as 'representing continuing lines of claims, as being, so to speak, deputies for a kind of everlasting moral agent or institution' (TJ, 128).

Rawls's theory assumes the structure of the family within the original position. But I would suggest instead that it be based on continuing political considerations which are openly debated in public forum. This strategy makes the representation of familial interest, a factor which strongly affects future generations, a subject of moral-political consideration. This is preferable to assuming a 'head of family,' whoever that might be, in an original position. This assumption might be dangerously exclusive, especially on a Kantian model where the concept of a person does not incorporate the effects of different family structures or the political-economic structure of a gender-segregated labor market.

The above argument raises the question of time-lapse and elicits a potential ambiguity on time in Rawls's original position. The original position, being hypothetical and only 'a device of representation' (JFP, 224), is outside of time or taken at an 'ideal point,' frozen in time (or metatime?). Strictly speaking, the original position is not formulated within the passage of historical time at all, although Rawls has recently noted that the conception of justice as fairness issues out of Western constitutional democracies in, presumably, the mid-twentieth century (JFP, 224).

Construction, on the other hand, is a process that must by definition occur over a temporal span. At least this is so on Dworkin's conception of construction as allowing for historical and cultural change. Rawls's notion of reflective equilibrium and recent emphasis on mid-twentieth century constitutional democracy implies at least some measure of temporal change within the process of construction. Certainly the model-conception of a well-ordered society lends itself more easily to this interpretation than the timelessness of the original position. The answers to these questions are crucial for consideration of how technology affects the cultural arena within which the principles of justice are formed. The effects of electronic communications media on democratic decision making and the effects of biotechnologies on reproductive and family social policy are two cases in point.

III. EQUALITY, NEUTRALITY, AND RESPONSIBILITY

Equality is one of the leading ideals of current constitutional democracy and is basic to Rawls's conception of justice as fairness. Rawls's original position is a way of representing equal concern and respect for everyone. Furthermore, in Rawls's conception of a well-ordered society the participants are modelled as free and equal citizens.

The ideal of equality, however, generates certain difficulties when we consider certain policy issues regarding future generations. These challenge the notion that equality must, so far as possible, be affirmed independently of a particular conception of a good life.[11] This 'neutrality principle' is affirmed, in different ways, by both Rawls and Dworkin. After investigating several challenges to a neutrality principle, I conclude by illustrating the importance of a constructivist model of morality or politics for addressing such challenges.

I argue here that a notion of equality which is neutral with regard to conceptions of the good life is not possible in certain social policy cases involving future generations.[12] The cases illustrate how a 'neutrality principle' for equality cannot be presumed apart from democratic process. It is, I argue, only possible as a process of open public discussion along lines of the accountability requirement in a constructivist theory. In other words, such problems cannot be solved without a theory of social and public responsibility.

Let us examine the way questions of equality might figure into questions of the good life by looking first at Ronald Dworkin's useful way of casting the question. He claims that equality is the central ideal and what he calls the 'nerve of liberalism' (L, 183). It is, along with the ideal of liberty, the 'constitutive political morality' of liberalism. Then, since 'each person follows a more-or-less articulate conception of what gives value to life,' the question arises whether political decisions and government generally can be neutral on the question of the good life.

Dworkin distinguishes two theories. One answers affirmatively, that 'political decisions must be, so far as possible, independent of any particular conception of the good life, or what gives value to life.' The reason is that since citizens 'differ in their conceptions, the government does not treat them as equals if it prefers one conception to another, either because the officials believe that one is intrinsically superior, or because one is held by the more numerous or powerful group' (L, 191).

The second theory holds on the contrary that 'the content of equal treatment cannot be independent of some theory about the good for humans or the good of life.' This is because 'treating a person as an equal means treating him the way the good or truly wise person would wish to be treated' (L, 191).

Dworkin argues that liberalism 'takes, as its constitutive political morality, the first conception of equality' (L, 192). How should we think of the kind of neutrality on the first concept? One answer, affirmed here, is that we should think of it as an achieved ideal within a constructivist model, rather than as an antecedently assumed prior notion whose content is invariant to historical and technological progress. A version of social policy making can be achieved through a public forum in which participants consider themselves responsible to each other and to the continuing forum.

To illustrate conflicting concepts of equality which are tied to a theory of the good life, consider a hypothetical example of state reproductive policy. In case A only the progeny of marriages legally accepted by the state are legally permitted to marry. Illegitimate offspring, or the children of non-sanctioned marriages, are not permitted to marry.[13] Let us suppose further that two distinct groups have opposing ideas of what to do about the proscription of marriages for illegitimates. Reformists argue that legitimacy laws undermine equal treatment for the offspring of non-sanctioned marriages. Further, they argue, such treatment imposes a conception of the good life by prescribing a procreative policy which excludes persons from full citizenship through no fault of their own. The value of life for these persons, argue the reformists, holds regardless of their origins.

In contrast, let us suppose that traditionalists who oppose change of legitimacy laws reply that extending marriage rights to illegitimates would encourage unregulated unions with definite negative effects on the well-being of the general population. If marriage regulations were modified, this group argues, the social fabric and unity of the society would be seriously damaged, and the equality of existing citizens undermined.[14]

Now Dworkin has said that political decisions must be, as far as possible, independent of any conception of the good life or of what gives value to life. To appeal to a purely procedural criterion of equality, on the one hand, will not support a neutral concept of equality in these examples. For the reformist position itself appeals to a notion of the value of each person, entitling that person to equal treatment regardless of parental status and parental wishes. The traditionalist position correspondingly appeals to a different conception of the good life for the community as a whole. Thus, a notion of equality which initially appeared purely procedural slides over into a notion of equality involving a conception of the good life whether we take the reformist or the traditionalist position. Hence, we could argue, there is no prior neutral notion of equality from which the two opposing standpoints may be evaluated. Nor is there a neutral concept which may be assumed in advance of the public debate and testing of the opposing positions.

In a second case B, suppose the existence of a cheap and easily applied genetic screening device for congenital abnormalities which severely impair infants at birth and continue to thwart and retard their physical and mental development when or if they survive into adulthood.[15] The policy question is, should such testing be publicized on a mass scale and made available by the state to all pregnant women at no charge? Should it be automatically administered unless citizens sign a dissent form?

View one answers affirmatively. Yes, it says, free testing should be publicized and available on a mass scale to support equal access to technologies of safe and healthy reproduction. If such testing were available only through private means, through private clinics, physicians and health plans, then economically privileged classes would probably avail themselves of it. Poorer groups, in contrast, might not even have knowledge of the technology. This would violate a principle of equality of access to the current resources of the society.

View two answers negatively. Testing on a mass scale, state-funded and widely publicized, could exert a coercive influence on pregnant women to utilize the screening technology, whereas such use should be founded on free and voluntary selection of the service. The wide availability of amniocentesis, for example, leads many women who would otherwise not consider the technology to feel constrained to use it. Such coercive influence interferes with full exercise of freedom of choice. Such mass screening also intrudes upon a sphere of private decision making and weakens the recognized U.S. constitutional principle of familial autonomy. A policy of mass testing at no charge should therefore not be adopted because it undermines equal rights of reproducers – in this case pregnant women – to decide the circumstances of their parenthood and the status of any progeny.[16]

Note that both views of equality consider only existing persons, not future individuals. Hence they avoid a paradox of future individuals.[17] That paradox might arise, however, if we considered a third view which rejected the testing because of an equal right to life of those potential persons, who, because of defects found by the screening technology, were aborted.

The differing notions of equality between views one and two come down to an opposition between equal access and the equal right to decide the circumstances and formation of one's family. The conflict occurs between an access and a right. This, however, is not the relevant consideration here as much as the conflict between two opposing emphases or priorities within the notion of equality itself. For consider a variant of view two which affirms the policy of mass testing on the basis of a person's equal right to decide the circumstances of one's family. The reasoning might be based on the fact that exercise of the right is enhanced by the greater latitude of choice and foreknowledge provided in the mass screening technology.

How would the question be decided, then, between the opposing notions of equality? Is either of the two views preferable because it is neutral respecting any ideal of the good life which might be involved? Is there any higher criterion which could decide? The answer to these two questions must be negative. In the first place, each of the views, one and two (but also three and the modified version of two), calls into play notions of equality which are inextricably tied to a notion of the good life. Even if the equal-access view one insisted that its emphasis was purely procedural, not tied to a conception of a good life, we could respond that its application did involve such a conception. In the social policy regarding reproductive biotechnologies, we cannot make a sharp separation between the principle of equal access to health resources and its application. Once we apply the policy of equal access, there is greater latitude for both choice, on view one, and coercion, on view two.

One way out of this problem may be to revert to Rawls's notion that in the original position the parties 'follow the same regulative conception of justice' (TJ, 582). Or that in the well-ordered society 'everyone accepts and knows that the others likewise accept the same principles' (KC, 537). However, reversion to some prior assumption of agreement will not take us out of the dilemma. In the above cases, each view does already accept the principles of equality regarding access and rights. The disagreement comes not from rejecting either of these principles but from placing different priorities on widely accepted elements within the idea of equality. Since the priority setting depends on a notion of what would make a better life for participants, it is not possible to establish a fully neutral concept of equality.

There is a further approach to the difficulty which proposes a social process rather than a resolution on the conceptual level. Instead of appealing to some neutral notion of equality we could appeal to the constructivist model of responsibility. Then, on the Integration Requirement (1) (see Section II, above), citizens and officials would be led to integrate the current technological proposal for screening with the prior judicial affirmation of fundamental right to liberty in reproductive decisions. The Consistency Requirement (2) would constrain them to harmonize the ideals in some compatible way. The Sincerity Requirement (3) demands openness about any eugenic motives. Finally, the Accountability Requirement (4) demands that the social policy decision making must be publicly debated and evaluated apart from special interests' consideration. The point is that it may be fruitless to search for a neutral concept here. The most we can expect is a democratic social process which directly addresses disparate views and hopefully arrives at some social consensus.

Another case presenting difficulties for a neutrality principle concerns medical technology and responsibility to succeeding generations. The controversial cases of seriously damaged newborns have been the topic of public and judicial disagreement in the mid-1980s. Suppose, for sake of simplicity, the following hypothetical arguments:

Group X argues that, regardless of parental circumstances and parental wishes, and even though the sustenance of life for seriously damaged newborns involves great social cost, these newborns have a right to equal treatment. They deserve, in other words, as much care and technological assistance as is available to sustain them, even if they are handicapped so seriously that they may die within a few years or months. On the basis of fair and equal treatment, and on pain of discrimination against the handicapped, they deserve this care. It is our responsibility to give it to them. Hence, argues group X, a decision making policy requiring care given to all handicapped newborns should be adopted.

Group Y claims that a policy of equal treatment across the board, neglecting a case by case decision making process enacted by parents, medical personnel, psychological support teams, makes a travesty of the ideal of equality. To provide equal treatment regardless of individual circumstances is in many cases simply to prolong and perhaps increase the suffering of such newborns. What we should do instead is place the decision making on a case by case basis and ensure that parents of all economic and educational backgrounds have equal access to medical care, should they decide to use it. An across the board policy makes the mistake of placing the highest priority on equal treatment at the expense of equal compassion for newborns who would otherwise soon die a natural death. The ideal of equality cannot be fairly applied without taking into consideration the quality of life in each case. To apply medical and technological intervention equally in every case is irresponsible with respect to the principle of respect for existing life of these newborns and may amount simply to prolonging death. A contextualized decision making process is therefore the best social policy to adopt in this case.[18]

In these two opposing views, as in the preceding case, what has been illustrated is that there is very little hope for a completely neutral concept of equality to adjudicate between these opposing conceptions of equality. It is doubtful that there is a higher-order notion of equality which is neutral regarding positions X and Y. Neutrality in such cases, in other words, cannot be assumed prior to either the judicial or the democratic process. If it occurs in such cases, it must be an end toward which we strive in a process rather than an ideal presumed independently of such process.

Equality, in these types of cases, is a functional ideal rather than an abstract, conceptual ideal. It must be enacted within democratic procedures. The constructivist model gives us some guidelines as to how this should happen. The canons for debate include the accountability requirement which 'does not allow appeals to unique intuitions that might mask prejudice or self-interest.'

Just as Rawls emphasizes the constructive process with the concepts of desert and freedom, so we could also emphasize it with the concept of equality. Desert, for example, is not a prior and independent notion that 'could override or restrict the agreement of the parties as agents of the construction' of justice as fairness (KC, 551). Citizens are free in the sense that they have the moral power to have a conception of the good. But this does not imply that they have a fixed conception, immune to revision. In fact Rawls stresses that 'as free persons, citizens claim the right to view their persons as independent from and as not identified with any particular conception of the good, or scheme of final ends' (JFF, 240).

Do these arguments for a constructivist model emphasizing democratic process imply that fundamental rights to liberty of procreation (which now exist according to U.S. constitutional precedent) – e.g., the right to contraception or abortion – should be erased and the question of their status brought back to legislative consideration or to referenda? I do not think this is implied by a constructivist model or by anything presented on the theory of responsibility. Simply to erase judicial precedent would undermine the integration and consistency requirements of the constructivist model. There is already, in our U.S. constitutional democracy, a commitment to a constitution and to its stated and implied rights. Hence to reject familial autonomy and liberty of procreation would be to fail on the consistency requirement. In other words, the judicial process itself, as Dworkin emphasizes, is subject to the same requirements he states in his constructivist model of a theory of responsibility.

It should be granted, however, that if substantial technological change placed these rights and liberties in a new light, they would have to come under both judicial re-evaluation and democratic debate. What I am thinking of is not simply the current controversy over abortion. Nothing said here undermines contraception and abortion rights, nor should it be presumed that the argument here implies that existence of these fundamental liberties should be subject to local decision making, any more than freedom of speech should. But if extrauterine gestation were to become technologically possible and genetic engineering a commonplace, surely we would review our concepts of fundamental rights and equality regarding procreation.[19]

In conclusion, on the topic of social and cultural conditions, I want to suggest that the development of biotechnologies makes social controls on human reproduction more explicit in social consciousness. They are sensitive moral and social questions because, in the course of things, each society must adopt an implicit or explicit social policy for dealing with marginal, damaged, or weaker offspring, or the potential to reproduce such offspring. Given the changing social mores and the plurality of current moral positions in contemporary Western democracies on these issues, it is even more important that responsibility to future generations involve the kind of democratic process sketched here in the theory of responsibility on a constructivist model.

Equality does have a place in these debates. Its ideal, however, involves not only equality of individuals but of groups, especially when we consider technological development. Does each group in the society receive equal treatment in medical technology and have free, informed access to technological controls on reproduction? I put emphasis on 'informed' because public education of both a technical and moral sort is required for the process sketched in Section I to succeed. In this way we avoid the dangerous consequence that some high authority in society, presumed to reason neutrally about equality, has superior insight into resolution of these problems. There is no pure Archimedean point of social and moral neutrality in the cases explored here. There is only the democratic forum in which we must participate as responsible citizens.

University of Massachusetts, Boston

NOTES

[1] Affirmation of rights is crucial, as I see it, for certain social issues, for such rights of the person as reproductive rights – the right to continue or terminate a pregnancy – and the right to refuse medical treatment, which are based on right to privacy in the U.S. Constitution precedent. Apart from this affirmation, there are problems in beginning with relying primarily on a rights framework in liberalism. For recent critiques, see Michael Sandel, *Liberalism and the Limits of Justice* (New York: Cambridge University Press, 1982); Charles Taylor, 'Atomism,' in Alkis Kontos, ed., *Powers Possessions and Freedoms: Essays in Honor of C. B. MacPherson* (Toronto: University of Toronto Press, 1979), pp. 39–61. Reprinted in *Philosophical Papers* (New York: Cambridge University Press, 1985). My arguments here for responsibility to future generations do not depend only on constitutional democracy and could be extended to other forms of government.

[2] 'Morality' shall be understood here to encompass politics.

[3] Abbreviations in the text to key works are as follows: Ronald Dworkin, *Taking Rights Seriously* (Cambridge: Mass.: Harvard University Press, 1977); chapter 5, 'Justice and Rights,' is cited as JR; R. Dworkin, 'Liberalism,' in *A Matter of Principle* (Cambridge, Mass.: Harvard University Press, 1985) is cited as L; John Rawls, *A Theory of Justice* (Cambridge, Mass.: Harvard University Press, 1971) is cited as TJ. 'Kantian Constructivism in Moral Theory: The Dewey Lectures,' *Journal of Philosophy*, 77 (September, 1980) is KC; and 'Justice as Fairness: Political Not Metaphysical,' *Philosophy and Public Affairs*, 14 (Summer, 1985), is JFP.

[4] Rawls apparently does not agree with Dworkin's observation in 'Justice and Rights' that the theory of TJ is rights-based as Dworkin characterizes it (KC, 236, note 19). This point, however, does not affect my argument here.

[5] Technological developments such as motorized chairs and voice systems have supported these changes.

[6] See Jurgen Habermas, *Communication and the Evolution of Society*, trans. Thomas McCarthy (Boston: Beacon Press, 1979), chapter 1, Universal Pragmatics, pp. 58–61; and *The Theory of Communicative Action*, vol. 1 of *Reason and the Rationalization of Society*, trans. Thomas McCarthy (Boston: Beacon Press, 1981), pp. 288–337. See also John Searle's 'sincerity rule,' in *Speech Acts* (London: Cambridge University Press, 1969).

[7] For one among many examples of feminist literature on these issues, see Michelle

Stanworth, ed., *Reproductive Technologies: Gender, Motherhood, and Medicine* (Minneapolis: University of Minnesota Press, 1987).

[8] Brian Barry, 'Justice Between Generations,' in P. Hacker and J. Raz, eds., *Law, Morality, and Society* (Oxford: The Clarendon Press, 1977); Derek Parfit, 'Future Generations, Further Problems,' and Gregory Kavka, 'The Paradox of Future Individuals,' *Philosophy and Public Affairs*, 11 (Spring,1982).

[9] My argument here appeals to the conception of a well-ordered society, not simply to Rawls's original position which I think incorporates different notions of both time and history.

[10] 'Justice Between Generations,' pp. 269, 271.

[11] The difficulty of giving arguments for substantive equality is discussed by T. M. Scanlon in 'Contractualism and Utilitarianism,' in A. Sen and B. Williams, eds., *Utilitarianism and Beyond* (Cambridge: Cambridge University Press, 1982).

[12] Compare the treatment by James Fishkin, *Justice, Equal Opportunity and the Family* (New Haven: Yale University Press, 1983), p. 158.

[13] There are historical examples of state prohibition of such marriages. Under Jewish religious law and Israeli state law today, a child born of an adulterous union, a 'Mamzer,' is 'forbidden to marry any other Jews except converts and other bastards. This ban is handed down for ten generations.' The dictum is a rabbinical decision and not made by a democratic legislature. See Lesley Hazelton, *Israeli Women* (New York: Simon and Schuster, 1977), p. 42.

[14] Such eugenic arguments have a recent historical precedent in the U.S. During the 1920s President Theodore Roosevelt joined the then-current tendency to speak of 'building the race.' He recommended six children for people of 'normal stock.' Those of 'better stock' should have more. 'Better stock' meant Yankee, i.e., non-immigrant populations. See T. Roosevelt, 'Race, Decadence,' *Outlook*, September 13, 1927, p. 111.

[15] See the discussion of techniques in H. Harris, *Prenatal Diagnosis and Selective Abortion* (Cambridge, Mass.: Harvard University Press, 1975). The techiques of alpha-feto protein analysis and chorionic villus biopsy are examples. The hypothetical cases might apply to the former in, for example, the state of California's recent pilot programs.

[16] The point of these examples is not whether one position or another is more cogent but the conflicting notions of the good life and of equality affirmed therein.

[17] See Gregory Kavka, 'The Pardox of Future Individuals,' *Philosophy and Public Affairs*, 11 (Spring, 1982).

[18] The newborn case obviously involves many issues, in particular, who should make the treatment decision, as well as familial right to privacy, federal vs. state vs. regulator agency jurisdiction, who is to be responsible for the care/cost of newborns. These cannot all be addressed here. They would, however, be addressed in social policy making by the integration requirement.

[19] This article focuses on developing a new theoretical framework for responsibility founded on egalitarian and democratic values. The focus on theory may understandably strike some readers as limited and incomplete, a criticism the author accepts and would remedy except for limitations on scope and space. Extension of the basic theory presented here includes possible programs for democratic *praxis* allied with a sense of responsibility as responsiveness to the needs of present and future generations. They

are explored in subsequent papers. The author considers them no less important despite the lack of space to explore them here.

NAME INDEX

Adam, J. A. 39
Adelman, Kenneth 13, 23
Adjukiewicz, Kazimierz 154, 165, 166
Aldridge, Robert 18, 19, 23, 24, 25
Aleksandrov, Alexei 23
Alhazen (Ibn Al-Haythan) 66
Allison, G. 23
Altszuller, G. S. 157
Antoniuk, G. A. 157, 166
Archer, B. 166
Arendt, Hannah 110
Aristotle 15, 63, 78, 98, 99, 110
Arkin, William 23
Arnheim, Rudolf 72, 75, 78

Babst, Dean 23
Bacon, Francis xiii, 86
Bacon, Roger 63, 66, 78, 104
Bailey, Lee W. 76, 78
Barfield, Owen 65, 66, 77, 78
Barnes, Barry 58
Barry, Brian 174, 175, 185
Bateson, Gregory 77
Beck, L. J. 104, 110
Bella, David 39, 40
Bentham, Jeremy 103
Bereanu, Bernard 5, 23
Bergson, Henri xvii
Berman, Morris 77, 78
Bernstein, Richard J. 93, 94
Bian Chunyuan 143, 148
Biden, Joseph R. 30
Blair, Bruce 19, 20, 25
Blaug, Mark 153, 154, 166
Boorstin, Daniel 101, 110
Borning, Alan 23
Boutwell, J. 23
Bowman, R. M. 39
Boydston, Jo Ann 93, 110

Bracken, Paul 22, 31, 39
Bradley, Morris 23
Brodsky, Garry 81
Bullert, Gary 94
Bunge, Mario 153, 160, 164, 165, 166
Burckhardt, Jacob 69, 78

Caesar, Julius 44
Camillo, Giulio 73, 74
Capra, Fritjof 77, 78
Carnesale, A. 23, 39
Carter, A. B. 39
Chen Bing 148
Chen Changshu 134, 136, 137, 138, 143, 144, 148
Chen Fan 137, 148
Chen Junhong 148
Chen Nianwen 143, 144, 148
Chen Wenhua 148
Chisholm, Roderick 115, 128
Cohen, D. 39
Colburn, Don 95
Copernicus, Nicholas 57, 99
Copp, D. 49, 58
Cragg, W. 58
Crandall, R. W. 42, 58
Craxi, Bettino 17, 24
Crissey, Brian 5, 7, 8, 9, 11, 23, 24, 39
Croke, K. G. 49, 59

D'Aguillon, Francois 75
da Vinci, Leonardo 65, 66, 67, 69, 75
Daguerre, Louis 63
Darwin, Charles 102
Della Porta, Giambattista 63, 71, 73, 74, 75, 78
Deng Honghai 139, 140, 148
Deng Shuzeng 148
Descartes, Rene xvi, 63, 64, 67, 68, 77, 78, 104

NAME INDEX

Dewey, John 81–90, 93, 94, 97–111, 123, 127, 130
Diao Peide 142, 148
Diderot, Denis 70, 71, 75
Donagan, Alan 14
Dong Shiyi 143, 148, 149
Durbin, Paul T. 110
Durer, Albrecht 71
Dworkin, Ronald 169–173, 177–179, 183, 184
Dykhuizen, George 93, 94

Edge, David 58
Ellsberg, Daniel 21
Ellul, Jacques xi
Everett, R. R. 40

Fan Hongye 149
Feuer, Lewis 81, 82, 93
Feuerbach, Ludwig 64, 76, 78
Finkelstein, A. C. W. 165, 166
Finkelstein, L, 165, 166
Finkin, Matthew W. 95
Firmage, Edwin B. 24
Fischetti, M. A. 39
Fishkin, James 185
Fletcher, J. R. 27, 28, 32, 34, 39
Fløistad, Guttorm xxiv
Fludd, Robert 73, 74
Foley, T. M. 40
Ford, Daniel, 23
Foucault, Michel 124, 129
Franklin, Benjamin xvii
Freud, Sigmund xiv, 64, 76, 77, 78
Friedlander, H. 23

Gadamer, H.-G. 114
Galileo 99, 100, 101, 104
Gao Dasheng, 145, 149
Gasparski, Wojciech 154, 164, 165, 166, 167
Gellner, Ernest 156, 165, 166
Gernsheim, Alison 63, 66, 67, 70, 78
Gernsheim, Helmut 63, 66, 67, 70, 78
Gillespie, B. 46
Goldstein, Robert A. 95
Gombrich, E. H. 71, 78
Goodman, Nelson 118–121, 123–129

Goscinski, J. 157, 166
Gottfried, Kurt 23
Grassi, Thomas 10
Gregory XIII 44, 45
Gregory, A. A. 166
Greve, Frank 23
Grinspoon, Lester 23
Grobstein, Clifford 42
Gu Zuxue 148
Guan Jintang 149
Guan Xipu 149

Habermas, Jurgen 114, 184
Hacker, P. 185
Hafner, D. 23
Haig, Alexander 21
Hampshire, Stuart 103, 110
Hao Zhensheng 141, 149
Harris, H. 185
Harris, John 78
Hazelton, Lesley 185
He Zhongxiu 149
Hecht, Hermann 76, 78
Hegel, G. W. F. xi
Heidegger, Martin xi, 77, 105, 111, 113, 114, 122–127, 129
Hempel, C. G. 129
Hewlett, Charles F. 94
Hickman, Larry 95
Hillman, James 65, 77, 79
Hobbes, Thomas xiv
Holmes, Oliver Wendell 171
Hong Xiaotao 149
Hood, Webster F. 110
Hook, Sidney 81
Horgan, J. 39
Hostelet, G. 158, 166
Houser, Gary 23
Hu Xiangming 149
Hua Daming 142, 143, 149
Hua Liguang 143, 144, 149
Huang Linchu 134, 137, 138, 139, 149
Hubbard, Ruth 174
Hutchins, Robert 108
Huygens, Constantijn 70, 79

Ihde, Don 110, 195

NAME INDEX

Jacobson, Robert L. 95
Jacques, R. 166
James, William 123
Jastrow, Robert 39
Jefferson, Thomas 103, 104
Ji Yuxing 149
Jin Guantao 141, 142, 149
Johnson, Clifford 23, 24
Jones, James 21

Kallen, Horace 93
Kang Rongping 142, 143, 149, 150
Kant, Immanuel 126
Kavka, Gregory 174, 185
Ke Liwen 139, 150
Keepin, W. 43, 58
Kepler, Johan 69, 99, 101, 104
Keyworth, George 28, 30, 39
Kidder, Ray 3
Kindschi, D. W. 23
Kircher, Athanasius 70
Kofman, Sarah 65, 66, 79
Kontos, Alkis 184
Kotarbinski, Tadeusz 153, 156, 157, 164, 166, 167
Kou Shiqi 150
Kozielecki, J. 166, 167
Krieger, David 23
Kripke, Z. 23
Kuhn, Thomas S. 44, 58, 124, 129

Ladd, E. C., Jr. 92, 95
Lave, L. B. 42, 58
LeBow, Richard N. 22, 23
Lee, Desmond 99
Legget, R. F. 44, 58
Leibniz, G. W. 115
Leiss, William 41
LeMay, Curtis E. 15, 16, 18, 21, 25
Lempinen, Edward 24
Levy, Edwin 49, 58
Lewis, David 129
Lewis, Flora 22
Li Huiguo 150
Li Pengcheng 150
Li Ziugo 142, 148
Lin Jian 141, 150
Lin Youji 150

Linstone, H. A. 160, 167
Lipset, S. M. 92, 95
Liroff, R. A. 58
Liu Dongzhen 138, 150
Liu Ji 142, 150
Liu Lixian 150
Liu Qingfeng 149
Liu Zeyuan 144, 145, 150
Lo Kuang 150
Locke, John 63, 64, 67, 68, 69, 75, 79
Long, F. A. 23
Lothstein, Arthur 88, 94
Lovejoy, Arthur O. 84
Lowrance, William 41, 45, 49, 58
Lu Pinyue 138, 150

MacLean, Douglas 24
March, Barbara 23
Margolis, Joseph 128, 129
Marx, Karl x, 65
Mattessich, R. 157, 167
McCray, L. 58
McMillan, B. 39
McNamara, Robert 20
Mele, Alfred 22, 24
Miller, D. 166, 167
Milton, S. 23
Moore, William 15
Morrison, David 23, 24
Moyers, Bill 24

Nadler, G. 157, 167
Nagel, Ernest 81
Newton, Isaac 104
Nietzsche, Friedrich 65
Notturno, Mark A. 128
Nowak, L. 158, 167
Nowotny, Helga 42, 58
Nunn, J. 36
Nunn, Sam 48
Nye, J. 23

Ortega y Gasset, Jose xvii, 107
Osborne, Woodley B. 95
Overton, W. S. 40

Pan Liangtao 150
Pan Shuming 150

Panofsky, Erwin 71, 79
Parfit, Derek 174, 185
Parnas, D. L. 39
Peirce, Charles Sanders 86, 113, 114, 123
Peng Jinan 151
Pepper, Stephen 63, 79
Perkins, John H. 57, 58
Perle, Richard N. 13, 30
Perrow, Charles 31, 39
Pinch, Trevor 51, 52, 59
Plantinga, Alvin 115–121, 123, 128, 129
Plato 20, 87, 98, 99, 101, 110
Polowinkin, A. E. 167
Powell, J. 166
Pringle, Peter 23
Pszczolowski, T. 166
Putnam, Hilary 129

Qin Qingwu 141, 150
Quine, W. V. 117–121, 123–129

Ravetz, Jerome R. 42, 59
Rawls, John 169, 170, 172–177, 180, 184, 185
Raz, Joseph 185
Reagan, Ronald 3, 28, 30
Richter, Jean 69, 79
Rickover, Hyman 4
Ridgeway, James 24
Risner, Friedrich 69
Roosevelt, Franklin D. 93
Roosevelt, Theodore 185
Rorty, Richard 63, 79, 128
Rosenberg, David 24
Rossi, Paolo 101, 110
Rousseau, Jean-Jacques 65
Russell, Bertrand 85
Ryle, Gilbert 64, 79

Sabra, A. I. 66, 79
Sagan, Carl 23
Saint-Simon, Louis x
Salter, Lawrence 41
Sandel, Michael 184
Sasser, James 18
Scalia, Antonin 19, 21

Scanlon, Thomas 185
Scheffler, Israel 129
Schott, Kaspar 70
Schwarz, Heinrich 70, 79
Searle, John 184
Sen, Amartya 185
Sennott, Linn 8, 23, 39
Shakespeare, William 73
Shen Mingxian 150
Shi Guozhu 150
Shi Yiqing 151
Simon, Herbert A. 153, 165, 167
Smith, Adam x
Smith, Dan 25
Smith, J. R. 40
Socrates 117
Solomon 73
Song Huichang 151
Sontag, Susan 70, 79
Sosigenes 44
Sovern, Michael I. 93, 95
Stanworth, Michelle 185
Steinbruner, John 3, 22, 39
Sterba, James 25
Su Ziyi 151
Sun Shuping 151
Swartzman, D. 49, 59

Taylor, Charles 184
Taylor, Harold 93
Thomas, Norman 81
Thompson, E. P. 25
Tondl, L. 167
Tribus, Myron 154, 167
Trotsky, Leon 81
Tsongas, Paul E. 30
Tuveson, Ernest 68, 79

Veblen, Thorstein 81, 83, 88
Vlcek, J. 167

Wakin, Mal 25
Waldrop, M. M. 39
Wallace, Michael 8, 23, 39
Wallich, P. 39
Wang Haishan 145, 151
Wang Hongbo 145, 151
Wasserstrom, Richard 25

NAME INDEX

Weber, Max xiii
Weinberg, Alvin 42, 59
Weinberger, Caspar 5, 18, 19, 21, 22, 24
Weinstein, Fred 23
Wessel, M. R. 42, 59
Whitehead, A. N. xv
Williams, Bernard 185
Winner, Langdon 40
Wojcicki, R. 165, 167
Woll, Matthew 84
Wood, Charles 23
Wu Yuanliang 150
Wylie, Congressman 5
Wynne, B. 42, 48, 58, 59

Xie Enze 151

Xie Shusen 148

Yan Kangnian 136, 137, 151
Yang Derong 138, 150, 151
Yates, Frances 72, 73, 79
Yuan Deyu 134, 136, 137, 138, 143, 144, 151

Zhai Zhihong 151
Zhang Jia 150
Zhang Letong 151
Zhang Xielong 145, 151
Zheng Shiming 151
Zou Shangang 136, 151
Zraket, C. A. 39
Zwicky, F. 157

PHILOSOPHY AND TECHNOLOGY

Series Editor: Paul T. Durbin

OFFICIAL PUBLICATIONS OF
THE SOCIETY FOR PHILOSOPHY AND TECHNOLOGY

---

1. *Philosophy and Technology*
   Edited by Paul T. Durbin and Friedrich Rapp.
   (Published as Volume 80 in 'Boston Studies in the Philosophy of Science')
   1983, xiv + 344pp. ISBN 90-277-1576-9
2. *Philosophy and Technology, II.* Information Technology and Computors in Theory and Practice.
   Edited by Carl Mitcham and Alois Huning.
   (Published as Volume 90 in 'Boston Studies in the Philosophy of Science')
   1986, xxii + 352pp. ISBN 90-277-1975-6
3. *Technology and Responsibility*
   Edited by Paul T. Durbin.
   1987, x + 392pp. ISBN 90-277-2415-6
4. *Technology and Contemporary Life*
   Edited by Paul T. Durbin.
   1988, viii + 320pp. ISBN 90-277-2570-5
5. *Technological Transformation.* Contextual and Conceptual Implications
   Edited by Edmund F. Byrne and Joseph C. Pitt.
   1989, xii + 314pp. ISBN 90-277-2826-7
6. *Philosophy of Technology.* Practical, Historical and Other Dimensions
   Edited by Paul T. Durbin.
   1989, xxiv + 192pp. ISBN 0-7923-0139-0

Kluwer Academic Publishers
DORDRECHT / BOSTON / LONDON